Urban Sustainability

The Urban Sustainability Book Series is a valuable resource for sustainability and urban-related education and research. It offers an inter-disciplinary platform covering all four areas of practice, policy, education, research, and their nexus. The publications in this series are related to critical areas of sustainability, urban studies, planning, and urban geography.

This book series aims to put together cutting-edge research findings linked to the overarching field of urban sustainability. The scope and nature of the topic are broad and interdisciplinary and bring together various associated disciplines from sustainable development, environmental sciences, urbanism, etc. With many advanced research findings in the field, there is a need to put together various discussions and contributions on specific sustainability fields, covering a good range of topics on sustainable development, sustainable urbanism, and urban sustainability. Despite the broad range of issues, we note the importance of practical and policy-oriented directions, extending the literature and directions and pathways towards achieving urban sustainability.

The series will appeal to urbanists, geographers, planners, engineers, architects, governmental authorities, policymakers, researchers of all levels, and to all of those interested in a wide-ranging overview of urban sustainability and its associated fields. The series includes monographs and edited volumes, covering a range of topics under the urban sustainability topic, which can also be used for teaching materials.

Yin Ma

Evaluation and Prediction Modeling of Territorial Space Use Efficiency

Theory, Construction, and Application

Springer

Yin Ma
China Aero Geophysical Survey
and Remote Sensing Center for Natural
Resources
China Geological Survey
Beijing, China

ISSN 2731-6483 ISSN 2731-6491 (electronic)
Urban Sustainability
ISBN 978-981-92-4395-2 ISBN 978-981-92-4396-9 (eBook)
https://doi.org/10.1007/978-981-92-4396-9

This Springer imprint is published by the registered company Springer Nature Singapore Pte Ltd.
The registered company address is: 152 Beach Road, #21-01/04 Gateway East, Singapore 189721, Singapore

To my beloved family, for their unwavering support, endless patience, and profound love throughout my academic journey.

To my supervisors and teachers, for their persistent guidance and inspiring insights throughout this research.

Preface

Territorial space serves as the spatial carrier for ecological civilization construction and the material foundation for high-quality economic and social development. Against the backdrop of deepening globalization and urbanization, territorial space use efficiency (TSUE) not only directly affects regional competitiveness but also determines the quality and resilience of a country's long-term development. At present, China is at a critical stage of promoting new-type urbanization and ecological civilization in a coordinated manner, and the reform of the national territorial spatial planning system has entered a deepening stage. How to scientifically and accurately evaluate and predict TSUE, break through resource and environmental constraints, and coordinate production-living-ecological space (PLES) has become a major issue of common concern in the fields of geography, urban planning, economics, and public administration.

However, the traditional research paradigm for TSUE is confronted with considerable limitations. Most mainstream evaluation and prediction models are based on the assumption of independent and identically distributed, which ignores the inherent heterogeneity and interdependence of geographic spaces. Consequently, evaluation results deviate from reality, and prediction accuracy cannot meet the needs of refined management. To tackle this "spatial blind spot", the academic community is in urgent need of a new methodology that can incorporate geospatial effects endogenously into the model framework. In view of this, this study takes the theory of geospatial effects as the core thread and the coordination of PLES as the application orientation and establishes a new framework for the evaluation and modeling of TSUE, covering indicator selection, efficiency measurement, and trend forecasting.

The core contribution of this study lies in its innovative methodology and modeling framework. First, this study overcomes the limitations of traditional factor analysis (FA) by developing a geographic factor analysis (GeoFA) model. Through a dual-path approach combining a geographically weighted correlation matrix and a spatial similarity matrix, it shifts the indicator selection process from a globally uniform approach to a locally adapted one. Second, this study introduces spatial constraints into data envelopment analysis (DEA), proposing the geographic data envelopment analysis (GeoDEA) model. By reconstructing the measure set through spatial clustering, it enhances the geographical rationality of the efficiency measure reference frontier. Third, this study integrates machine learning algorithms with cellular automata (CA) model to construct an extended CA model. This transforms the forecasting paradigm from experience-driven to data-driven, significantly improving the accuracy and robustness of long-term trend simulations.

In terms of empirical analysis, this study focuses on 340 prefecture-level and above cities in China. It conducts a long-term spatiotemporal evolution analysis covering the period from 1990 to 2020 and provides scenario projections for 2030. The study indicates that TSUE in Chinese cities exhibits significant spatiotemporal heterogeneity and spatial agglomeration characteristics, with 2010 marking a critical turning point in its evolution. Production space use efficiency (PSUE) dominates the changes in overall efficiency, while living space use efficiency (LSUE) and ecological space use efficiency (ESUE) show distinctive patterns of regional differentiation. Based on precise GeoDEA model measurements and extended CA model dynamic simulations, this study further proposes a policy framework categorized by dimensions, regions, and types. It provides a scientific basis and decision-making support for territorial spatial governance in the new era.

This study embodies an in-depth exploration of interdisciplinary integration, dedicated to forging a rigorous connection between geospatial analysis and efficiency evaluation. The theoretical framework, model construction, and empirical examinations elaborated in the study are all grounded in the author team's sustained and systematic research. As such, this study constitutes a valuable reference for scholars and practitioners in the fields of territorial spatial planning, resource management, and urban studies, while also offering robust decision support for relevant government authorities and policy-making bodies.

Beijing, China Yin Ma

Acknowledgments Upon the completion of this monograph, I would like to extend my sincerest gratitude to all mentors, colleagues, relatives, and friends who have offered generous help and persistent support throughout this research.

I am particularly indebted to Professor Zheng Xinqi from China University of Geosciences, Beijing. His rigorous guidance and forward-looking insights have guided the entire process of this study, ensuring the smooth progress and final completion of this monograph.

I sincerely appreciate my colleagues and peers for their valuable academic discussions and practical assistance in model development, empirical analysis, and data processing throughout the research.

I also extend my deepest thanks to my family and friends for their consistent companionship, tolerance, and encouragement, which have sustained my academic perseverance along this journey.

All remaining deficiencies and errors in this monograph are solely my own responsibility.

Competing Interests The author has no competing interests to declare that are relevant to the content of this manuscript.

Contents

1 The Necessity and Practical Challenges of Territorial Space Use Efficiency Evaluation 1
1.1 Historical Context and Strategic Significance of Efficiency Evaluation 1
1.1.1 Spatial Resource Constraints under Globalization and Urbanization 1
1.1.2 Dual Demands of China's New-Type Urbanization and Ecological Civilization 2
1.1.3 Driving Effects of National Spatial Planning Reform on TSUE Evaluation 3
1.1.4 Contemporary Needs for PLES Synergy and Urban Governance Modernization 3
1.2 Conceptual Definitions and Theoretical Framework 4
1.2.1 Definitions of Core Concepts 4
1.2.2 Core Theoretical Foundations 5
1.2.3 Theoretical Connotation and Logical Relevance of PLES 7
1.3 Practical Dilemmas and Research Gaps in Efficiency Evaluation 11
1.3.1 Theoretical Dilemmas: Multidisciplinary Complexity and Inadequate PLES Coordination 11
1.3.2 Methodological Dilemmas: Bottlenecks in Geospatial Measurement and Prediction Modeling 12
1.3.3 Application Dilemmas: Challenges in PLES Differentiated Evaluation and Targeted Policy Regulation 13
1.4 Research Objectives, Core Content, and Scope Definition 13
1.4.1 Research Objectives 13

1.4.2 Core Research Content ... 14
1.5 Technical Roadmap ... 15
1.6 Core Innovations ... 18
References ... 18

2 Evolution of Models for Territorial Space Use Efficiency Research ... 21
2.1 Origin and Evolution of Territorial Space Use Efficiency ... 21
2.2 Research on Indicator System Design Models ... 23
2.3 Research on Efficiency Measurement Models ... 25
2.4 Research on Efficiency Prediction Models ... 29
2.5 Chapter Summary ... 30
References ... 31

3 Indicator Selection: Construction of Geographic Factor Analysis Model ... 35
3.1 Model Evolution and Theoretical Basis: From Traditional Factor Analysis Model to Geographic Factor Analysis Model ... 35
3.1.1 Core Logic and Spatial Limitations of the Traditional FA Model ... 35
3.1.2 Theoretical Connotation of Geospatial Effects: Spatial Dependence and Spatial Heterogeneity ... 37
3.1.3 Theoretical Challenges and Adaptation Requirements of Spatial Non-stationarity Theory for Traditional Models ... 38
3.1.4 Design Concepts of the GeoFA Model: From Global Uniformity to Local Spatial Adaptation ... 39
3.2 Dual-path Modeling for Model Transition: Geographically Weighted and Spatial Similarity Matrices ... 40
3.2.1 Benchmark Model: Mathematical Formulation and Applicable Conditions of the Traditional FA Model ... 40
3.2.2 Path 1: GeoFA Model with Geographically Weighted Correlation Coefficient Matrix ... 42
3.2.3 Path 2: GeoFA Model with Geographically Weighted Spatial Similarity Matrix ... 47
3.3 Model Comparison and Empirical Selection ... 53
3.3.1 Comparative Experiment Design ... 53
3.3.2 Analysis of Empirical Results ... 55
3.3.3 Comprehensive Conclusions ... 57
3.4 Chapter Summary ... 60
3.4.1 Methodological Contributions: A Full Evaluation Framework for the GeoFA Model ... 60
3.4.2 Applicable Scope and Notes on GeoFA Model ... 61
3.4.3 Future Outlook: From GeoFA Model to Complex Spatial Structure Modeling ... 61
References ... 62

4 Efficiency Measurement: Construction of Geographic Data Envelopment Analysis Model ... 63
4.1 Model Evolution and Theoretical Basis: From Traditional Data Envelopment Analysis to Geographic Data Envelopment Analysis Model ... 63
4.1.1 Traditional DEA Model: Basic Principles and Hypothetical Limitations ... 63
4.1.2 Geospatial Effects: Spatial Heterogeneity and Dependence in Efficiency Measurement ... 67
4.1.3 Core Idea of GeoDEA Model: An Efficiency Evaluation Framework under Spatial Constraints ... 68
4.2 Geographic Data Envelopment Analysis: Spatial Constraints and Measurement Reconstruction ... 69
4.2.1 Model Construction Mechanism: Reconstruction of Measurement Sets and Introduction of Generalized DEA Model ... 69
4.2.2 Measurement Reconstruction: Division of DMU Sets Via Spatially Constrained Multi-Dimensional Clustering ... 70
4.2.3 Model Coupling Mechanism: Integration of Spatial Clustering Model and Generalized DEA Model ... 74
4.3 Implementation and Verification of Geographic Data Envelopment Analysis: A Complete Methodological Framework ... 76
4.3.1 Design of Comparative Model Experiments ... 76
4.3.2 Model Rationality Verification: A Multi-dimensional Inspection Framework ... 77
4.3.3 Comprehensive Conclusions ... 92
4.4 Chapter Summary ... 93
4.4.1 Core Characteristics and Methodological Contributions of GeoDEA Model ... 93
4.4.2 Applicable Scenarios and Application Notes of GeoDEA Model ... 94
4.4.3 Limitations and Future Research Prospects of GeoDEA Model ... 95
References ... 96

5 Trend Prediction: Construction of Extended Cellular Automata Model ... 97
5.1 From Traditional Cellular Automata Model to Extended Cellular Automata Model: The Evolution of Simulation Paradigm ... 97
5.1.1 Traditional CA Model: Basic Principles and Limitations in Simulation ... 97

5.1.2 Machine Learning: Technical Challenges and Innovations for Simulation Optimization 98
5.1.3 Core Ideas of Extended CA Model: A Framework Combining Data Driving and Knowledge Guidance 100
5.2 Theoretical Basis: Overview of Core Model Principles 101
5.2.1 Spatial Feature Extraction: Fundamentals of CNN Model 101
5.2.2 Time-Series Dependency Modeling: Fundamentals of LSTM Model 102
5.2.3 Nonlinear Relationship Fitting: Fundamentals of RF Model 104
5.2.4 Dynamic Simulation Engine: Principles and Classical Structure of CA Model 105
5.3 Design of Extended Cellular Automata Model: Integrating Machine Learning into Traditional Cellular Automata Model 108
5.3.1 Model Coupling Idea: From Rule-Driven to Data-Driven Mode under Rule Constraints 108
5.3.2 Model Coupling Mechanism: Multi-level Modular Fusion Architecture 109
5.4 Implementation and Validation of Extended Cellular Automata Model 115
5.4.1 Design and Implementation of Model Validation Experiments 115
5.4.2 Data Preprocessing and Parameter Configuration 116
5.4.3 Model Validation: Multi-Dimensional Evaluation of the Coupled Model 121
5.4.4 Comprehensive Conclusions 123
5.5 Chapter Summary 123
5.5.1 Characteristics and Contributions of the Extended CA Model 123
5.5.2 Application Scenarios and Notes of the Extended CA Model 125
5.5.3 Limitations and Future Research Directions of the Extended CA Model 126
References 127

6 Empirical Analysis: Multi-dimensional Evaluation and Spatiotemporal Variation of Territorial Space Use Efficiency in Chinese Cities 131
6.1 Research Basis: Study Area, Research Data, and Analytical Framework 131
6.1.1 Overview of the Study Area 131
6.1.2 Data Sources and Preprocessing 133

6.1.3 Overall Analytical Framework: From Efficiency Measurement to Spatiotemporal Analysis and Prediction 136
6.2 Construction of Evaluation Indicator System for Territorial Space Use Efficiency 138
6.2.1 Construction of PSUE Indicator System 138
6.2.2 Construction of LSUE Indicator System 147
6.2.3 Construction of ESUE Indicator System 156
6.2.4 Dimensional Integration for Comprehensive TSUE Calculation 164
6.3 Spatiotemporal Evolution and Spatial Autocorrelation Analysis of Territorial Space Use Efficiency 165
6.3.1 Spatiotemporal Evolution of TSUE 165
6.3.2 Spatial Autocorrelation of TSUE 176
6.4 Prediction and Analysis of Territorial Space Use Efficiency in 2030 188
6.4.1 Temporal Variation of TSUE in 2030 188
6.4.2 Spatial Pattern of TSUE in 2030 189
6.5 Chapter Summary 191
6.5.1 Key Findings: National TSUE Features and Evolution Trends 191
6.5.2 Framework Discussion: Practical Value of the Analytical System 193
References 194

7 Policy Recommendations for Improving China's Territorial Space Use Efficiency 195
7.1 Theoretical Foundation: From Efficiency Measurement to Spatial Governance 195
7.2 Multidimensional Optimization Paths: Differentiated Strategies for Production-Living-Ecological Spaces 196
7.3 Zonal Classified Policy Framework: Governance Strategies for Spatial Efficiency Patterns 197
7.4 Institutional Safeguards and Policy Tool Innovation 198
7.5 Chapter Summary 200

Abbreviations

CA	Cellular Automata
CNN	Convolutional Neural Network
DEA	Data Envelopment Analysis
DEM	Digital Elevation Model
DMU	Decision-Making Unit
ESUE	Ecological Space Use Efficiency
FA	Factor Analysis
GeoDEA	Geographic Data Envelopment Analysis
GeoFA	Geographic Factor Analysis
LSTM	Long Short-Term Memory
LSUE	Living Space Use Efficiency
PLES	Production-Living-Ecological Space
PSUE	Production Space Use Efficiency
RF	Random Forest
RNN	Recurrent Neural Networks
SBM	Slack-Based Measure
SFA	Stochastic Frontier Analysis
TSUE	Territorial Space Use Efficiency

Chapter 1
The Necessity and Practical Challenges of Territorial Space Use Efficiency Evaluation

Abstract This chapter examines the necessity and practical challenges of evaluating territorial space use efficiency (TSUE) in the context of globalization, new-type urbanization, and ecological civilization construction. It clarifies the strategic value of efficiency evaluation under spatial resource constraints and amid the reform of the national territorial spatial planning system. Drawing on core theories (including geospatial effects, input-output analysis, and production frontier theory), the chapter further defines the connotations of TSUE and of production-living-ecological space (PLES). The chapter then dissects research dilemmas from theoretical, methodological, and application perspectives and specifies the research objectives, core contents, spatiotemporal boundaries, and technical roadmap. Finally, it summarizes the study's innovations in methodology, model system, and practical application, thereby establishing a solid theoretical foundation and logical framework for the subsequent model construction and empirical research.

1.1 Historical Context and Strategic Significance of Efficiency Evaluation

1.1.1 Spatial Resource Constraints under Globalization and Urbanization

The deep advancement of globalization and urbanization has driven the rapid agglomeration of production factors including population, capital, and technology in urban areas, continuously raising the intensity of territorial space development. However, this process has also intensified the imbalance between the supply and demand of spatial resources, resulting in prominent resource constraints. From a global perspective, urban expansion is characterized by the coexistence of horizontal sprawl and three-dimensional growth. The scattered layout in some cities has led to inefficient spatial utilization, exacerbating the waste of land resources and pressure on the ecological environment. At present, global spatial resources are limited in total volume and unevenly distributed. The rigid demand for spatial resources driven by

Y. Ma, *Evaluation and Prediction Modeling of Territorial Space Use Efficiency*, Urban Sustainability, https://doi.org/10.1007/978-981-92-4396-9_1

population growth and economic development stands in sharp contradiction with the scarcity and non-renewability of such resources [1].

Meanwhile, regional competition has become increasingly fierce under globalization, and the efficiency of spatial resource utilization has become one of the core indicators for evaluating regional competitiveness. The traditional extensive model of spatial development has not only led to the inefficient use of land and ecological resources but also restricted regional sustainable development. In China, more than 300 cities serve as the core carriers of urbanization. Amid the wave of globalization and urbanization, these cities are faced with the common challenges, among which the shortage of spatial resources, unbalanced utilization, and prominent spatial effects are particularly severe. Therefore, carrying out evaluations of TSUE and resolving the constraints of spatial resources is an inevitable choice for achieving high-quality development in the process of globalization. It also constitutes an important regional practice that addresses the pain points of spatial utilization in Chinese cities and responds to the needs of global spatial governance [2].

1.1.2 Dual Demands of China's New-Type Urbanization and Ecological Civilization

Against the macro backdrop of spatial resource constraints stemming from globalization and urbanization, China is at a critical stage of synergistically promoting new-type urbanization and ecological civilization construction [2]. These dual strategic objectives place higher demands on TSUE and constitute an important practical foundation for conducting efficiency evaluations. With the core principles of intensification, efficiency, livability, and appropriateness, new-type urbanization rejects the traditional extensive model of spatial development. It emphasizes optimizing spatial layouts and improving utilization efficiency, driving the transformation of urban development from scale expansion to quality enhancement. The evaluation of TSUE serves as a vital instrument for assessing the quality of urbanization and optimizing the allocation of spatial resources. Concurrently, ecological civilization construction requires strict observance of the ecological protection red line and integrated planning of production, living, and ecological spaces, so as to embody the development philosophy that "lucid waters and lush mountains are invaluable assets" [3]. This necessitates regulating the development intensity of ecological spaces and preserving the functional integrity of ecosystems through scientific efficiency evaluation. New-type urbanization and ecological civilization construction are complementary and dialectically unified. The former is underpinned by efficient spatial utilization, while the latter takes scientific efficiency evaluation as an effective approach. These trends compel Chinese cities to break away from the traditional spatial utilization model and realize the coordinated and efficient use of production, living, and ecological spaces [4].

1.1.3 Driving Effects of National Spatial Planning Reform on TSUE Evaluation

Based on the dual demands of China's new-type urbanization and ecological civilization construction, the state has successively issued relevant policies for guidance. Among these, the reform of the territorial spatial planning system has become the core driving force for promoting the evaluation of TSUE. In 2019, the Central Committee of the Communist Party of China and the State Council issued Several Opinions on Establishing the Territorial Spatial Planning System and Supervising Its Implementation. This marks a profound transformation of China's spatial planning from conflicting multiple plans to the integration of multiple plans into one. The new territorial spatial planning system is centered on the system for the development and protection of territorial space, and attaches equal importance to bottom-line constraints and efficiency improvement [5]. The system requires the scientific optimization of the territorial spatial pattern on the basis of the evaluation of resource and environmental carrying capacity and the evaluation of territorial spatial development suitability. This round of reform has realized the integration and unification of multiple plans, including major function-oriented zone planning, land use planning, and urban-rural planning. It has further clarified the layout orientation as well as the development and protection requirements for production, living, and ecological spaces. It has also provided a unified planning basis and standard framework for the evaluation of TSUE. Meanwhile, the reform of the planning system has emphasized the strengthening of territorial spatial monitoring, early warning, and performance assessment. It proposes to establish a scientific efficiency evaluation mechanism for the dynamic monitoring and accurate assessment of spatial use efficiency. The core implementation context of this requirement is at the city level. Chinese cities are key units for the implementation of territorial spatial planning. The evaluation of their spatial use efficiency must fully consider geospatial heterogeneity, so as to provide accurate data support and a decision-making basis for planning formulation and use control. This further highlights the practical significance of constructing a geospatial effect model in this study, forming a natural connection with the research focus [6].

1.1.4 Contemporary Needs for PLES Synergy and Urban Governance Modernization

Against the backdrop of globalization and the dual imperatives of new-type urbanization and ecological civilization construction, this study focuses on urban contexts and addresses its own core objectives [7]. China's urban spatial governance is currently confronted with prominent practical dilemmas, which constitute the fundamental starting point of this research. In urban spatial governance, the imbalance between production, living, and ecological spaces has become increasingly conspicuous [8]. Some cities prioritize the development of production spaces while neglecting ecological protection and the improvement of living quality, resulting in low spatial use efficiency. Most existing evaluations of TSUE ignore geospatial heterogeneity and spatial dependence. Such approaches fail to accurately characterize efficiency differences across cities and spatial types and cannot meet the requirements of modernized spatial governance [9]. In light of the above-mentioned background, demands and policy orientations, two major problems urgently need to be solved: the imbalance of the three types of urban spaces and the insufficient accuracy of efficiency evaluation. Accordingly, this study aims to construct an evaluation and prediction model for TSUE that takes into account the three types of spaces and highlights geospatial effects. This model can provide scientific support for the modernization of urban spatial governance. It reflects both the contextual background and core value of this study, ultimately forming a complete logical chain of "background–demand–policy–research core".

1.2 Conceptual Definitions and Theoretical Framework

1.2.1 Definitions of Core Concepts

1.2.1.1 TSUE

According to research, the concept of efficiency originated in mechanical engineering. It is defined as the ratio of useful power to driving power, measuring the effectiveness of mechanical work. This definition falls within the scope of physics. With the development of multiple disciplines, efficiency has acquired diverse connotations. In management studies, efficiency refers to the workload accomplished per unit time, or the effective allocation of social resources to satisfy human needs. Economists have introduced efficiency into economics to describe the input-output relationship of production factors in economic activities [10, 11]. Territorial space use refers to a series of production and living activities based on various resources within a given territorial space. Such activities possess economic, social, and geographical attributes. Combining these attributes with the above interpretations of efficiency, TSUE can be defined as the ratio between the total input of land, socio-economic, ecological, and other resources, and the total output of spatial, social, and ecological benefits within a specified territorial space.

1.2.1.2 PLES

The 18th National Congress of the Communist Party of China put forward for the first time the requirement of "promoting intensive and efficient production space, livable and appropriate living space, and ecological space with lucid waters and lush mountains". This formulation laid the foundation for the conceptual definition of production, living, and ecological spaces. Production, living, and ecological spaces are the collective term for production space, living space, and ecological space. These three types of spaces jointly constitute the core functional elements of territorial space across different scales [12]. Production space is a functional area that provides material resources for human survival and development. It is primarily dedicated to productive activities such as grain production, industrial manufacturing, and commercial services, and mainly includes construction land and agricultural land. Living space is a functional area that hosts human activities including residence, consumption, education, medical care, and recreation. Its core functions are rest, social interaction, and physical and mental well-being; it mainly includes urban residential areas, rural settlements, land for education and science, and land for medical and health care. Ecological space is a territorial area that provides ecosystem services and safeguards ecological security. Its main land use types include forest land, grassland, park green space, and water areas [13].

1.2.2 Core Theoretical Foundations

1.2.2.1 Geospatial Effect Theory

Geospatial effects refer to the impacts on geographic analysis results caused by the geographical characteristics, spatial distribution, and derived spatial allocation of various elements in a geographic space. Some scholars define geospatial effects differently [14]. They view them as complex phenomena in the transmission of matter, energy, and information within a geographic system. These phenomena arise from spatial factors such as differences in surface structure and changes in spatial pattern. These factors will further influence the outcomes of geospatial analysis. As scholars' understanding of geospatial effects has deepened, related research has been continuously expanded and refined. At present, geospatial effects can be summarized into four categories: spatial heterogeneity, spatial proximity, distance decay, and spatial scale [14, 15]. These four effects follow a logically progressive relationship, as shown in Fig. 1.1. Among them, spatial heterogeneity is the most fundamental effect. It provides a basis for the formation of other effects. Based on spatial heterogeneity, spatial proximity and distance decay effects have gradually emerged. These two effects reflect the second-order relationships of geospatial locations. Meanwhile, through aggregation and refinement of regional units, the spatial scale effect has been further developed. At present, many spatial analysis methods have fully incorporated geospatial effects. This ensures that geospatial analysis is conducted reasonably, accurately, and scientifically. Typical examples include the geographically weighted regression model, spatial autocorrelation model, spatial autoregressive model, spatial error model, and spatial Durbin model. These models have been modified and optimized by considering geospatial effects. Consequently, the accuracy and reliability of geospatial analysis have been significantly improved.

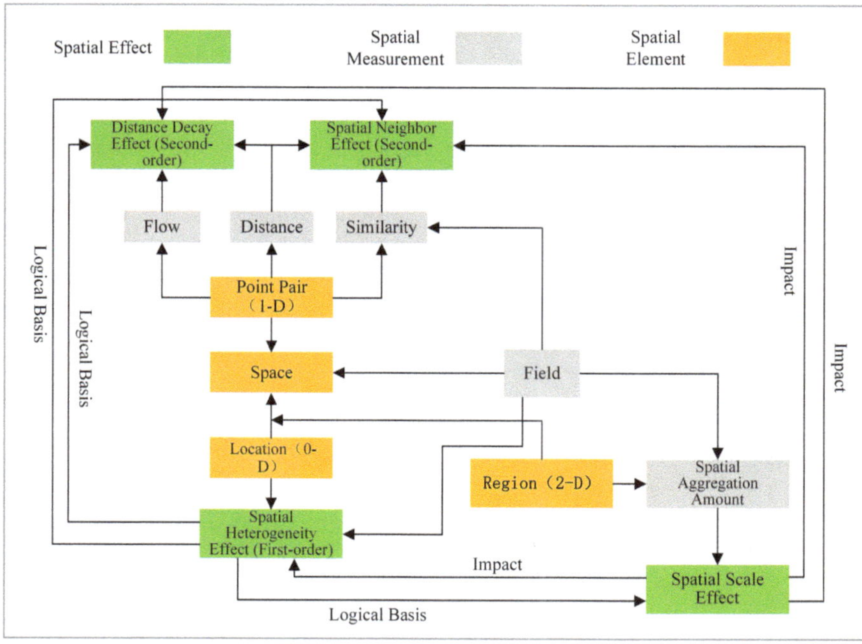

Fig. 1.1 Logical relationship among the four types of geospatial effects[1]

1.2.2.2 Input-Output Theoretical Connotation

Input-output theory describes the balance between production inputs and product distribution across various sectors of a national economy. It is rooted in Walrasian general equilibrium theory and serves as a quantitative analytical method in economics [16, 17]. The intellectual origin of this theory can be traced to Tableau Économique by François Quesnay. It was further developed in subsequent economic studies. In 1936, Wassily Leontief formally proposed the theory in Quantitative Input-Output Relations in the Economic System of the United States. He provided the first formal definition of input-output, marking a new stage in the development of the theory [18]. Through practical research by scholars across multiple disciplines, the input-output theoretical system has gradually matured and improved. It has become widely appreciated for its clarity and operability. The theory mainly explores quantitative relationships within complex economic systems by constructing mathematical equation models. It formalizes the mathematical relationship between input resources and output values of economic activities as functional equations. Optimal solutions are then derived to quantitatively measure efficiency in economic activities. From an efficiency-oriented perspective, input-output theory focuses not only on the

[1] The figure is adapted from the work of Liu et al. [14].

ratio of inputs to outputs. It also emphasizes how to adjust and optimize this relationship under given technological conditions to maximize the input-output ratio. In addition, a wide variety of models have been developed based on input-output theory. These include static input-output models, dynamic input-output models, and monetary input-output models. Such models have been widely applied in economics, medicine, geography, earth sciences, energy science, and other related fields [19, 20].

1.2.2.3 Production Frontier Theory

In the production theory of economics, production functions are commonly used to describe technological relationships in production. Under given production technologies, they represent the maximum output achievable from various factor inputs. The production possibility frontier described by such functions is known as the production frontier [19]. In 1957, the economist Michael Farrell conducted the first systematic research on the production frontier. Based on the idea of measuring production efficiency, his pioneering work opened a new chapter in the theoretical study of the production frontier [21]. Farrell's modeling approach for the production frontier involves selecting appropriate input and output variables and constructing a linear programming model to estimate a convex boundary. This boundary is treated as the production frontier. The distance of each decision-making unit (DMU) from the frontier is then measured. Missing or excessive inputs are identified and adjusted accordingly, so that targeted optimization suggestions can be provided to improve efficiency [22]. Specifically, if the output of a DMU lies on the production frontier, its production activity is regarded as technically efficient. If not, it has not achieved full efficiency under the existing technology, indicating either insufficient inputs or wasteful overuse of resources. At present, the production frontier theory has been widely applied in many fields, including enterprise management, land use, industrial performance, and agricultural production. In terms of research methods, the theory has developed two major branches: parametric analysis and non-parametric analysis, each with its own advantages and limitations [23].

1.2.3 Theoretical Connotation and Logical Relevance of PLES

1.2.3.1 Composition and Functional Characteristics of the Territorial Space Use System

The territorial space use system is a complex dynamic system that couples two core components: natural and human systems [24]. Internal elements of the system are interwoven, interactive, and mutually restrictive. Nonlinear relationships of competition and coordination exist between pairs of elements. Complex interactions among these elements result in multiple characteristics of the structure and function of the territorial space use system. These include systematic differentiation, functional complexity, and spatiotemporal heterogeneity [25]. Therefore, identifying and clarifying the structure and function of the territorial space use system is crucial for research on TSUE. Scholars worldwide have conducted in-depth investigations on this topic. Overall, classifying territorial space use based on its primary and secondary functions has become a preferred approach [24, 25]. Guided by national policy priorities, this study subdivides the territorial space use system according to its production, living, and ecological functions. It is refined into the PLES use system, including production space use, living space use, and ecological space use subsystems, as illustrated in Fig. 1.2.

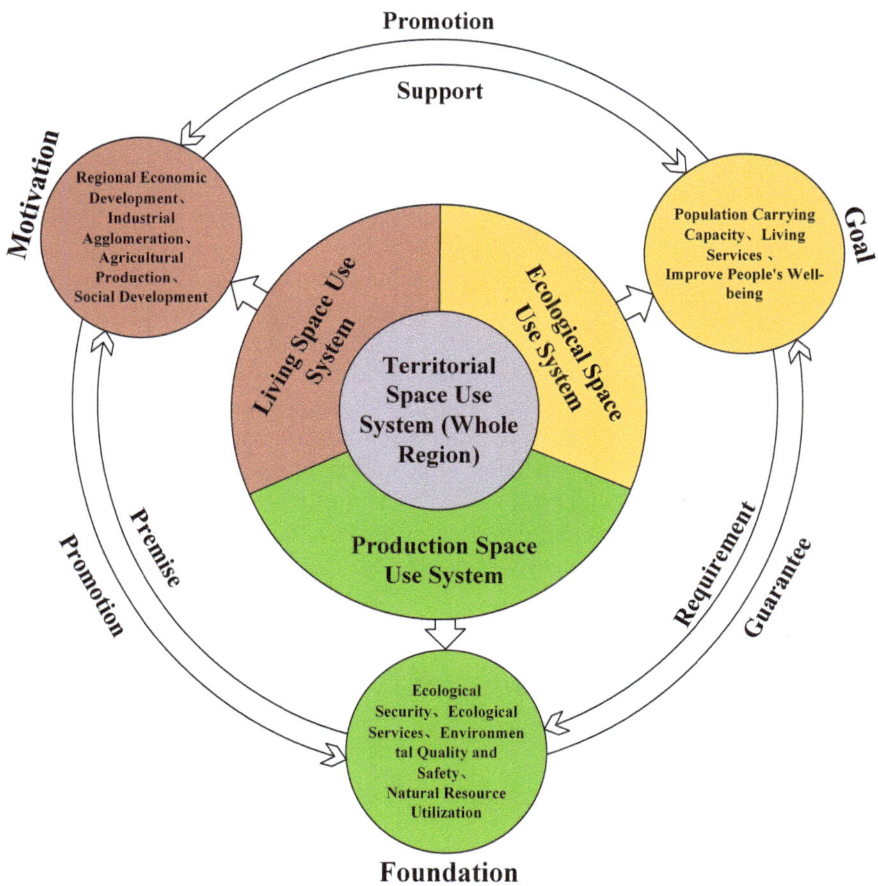

Fig. 1.2 Structure and function of the territorial space use system

1. **Production Space Use System**

   As shown in Fig. 1.2, the production space use system serves as the driving force of the territorial space use system. Its production function refers to activities carried out in production spaces, which directly or indirectly provide various products and services. This system undertakes multiple roles, including regional economic development, industrial agglomeration, agricultural production, rural culture inheritance, product supply, and stable social development. First, the system is mainly composed of cultivated land, grassland, water areas, as well as urban-rural construction land, industrial and mining land, and residential land. It acts as a key carrier supporting the operation of primary, secondary, and tertiary industries and plays a vital role in economic development. For example, cultivated land not only provides people with grain, oil, vegetables, and other daily necessities, but also serves as an important carrier for inheriting distinctive rural culture and enriching public life. Construction land, including urban-rural, industrial and mining, and residential land, supports production construction, industrial agglomeration, and professional division of labor. It exerts a profound influence on surrounding economic development with remarkable driving and radiation effects, mainly supplying industrial products and services for human society [26, 27]. In addition, the sound development of the production space use system provides important support for the improvement of living spaces. This support is reflected in product supply, service quality, cultural inheritance, and other aspects. Meanwhile, the coordinated development of production spaces also promotes the rational use of natural resources and the protection of the ecological environment.
2. **Living Space Use System**

   As shown in Fig. 1.2, the living space use system represents the core goal of the territorial space use system. Its living function mainly lies in providing settlements for human survival and development, and satisfying people's material and spiritual needs. Specifically, residential land provides places for human habitation. Commercial and service land meets daily consumption, entertainment, and leisure needs. Land for education supports learning and personal development, while medical and health care land plays a key role in safeguarding public health [28]. In addition, the living space use system influences industrial restructuring in production spaces and provides strong support for industrial development. It also puts forward clear requirements for the ecosystem services supplied by ecological spaces. This helps promote the optimization of land use structure in ecological spaces to better meet human survival needs.
3. **Ecological Space Use System**

   As shown in Fig. 1.2, the ecological space use system serves as the foundation of the territorial space use system. It is an indispensable part of territorial space. The ecological function of this system refers to the supply of necessary ecological products and services for human production and living during the operation of ecosystems. This system includes a variety of important functions, such as ecological service provision, water conservation, biodiversity maintenance, soil and water conservation, windbreak and sand fixation, and coastal ecological stability protection [29, 30]. The ecological space use system is mainly composed of forest land, grassland, water areas, cultivated land, marine areas, and unused land. These land use types are interwoven to form a green ecological network. It plays an important regulatory role in ecological security related to the atmospheric environment, biological communities, soil and water conservation, and other aspects. In addition, the abundant natural resources in the ecological space use system provide fundamental support for the sound development of production spaces. Meanwhile, various ecological services supplied by ecological spaces also support the healthy development of living spaces.

1.2.3.2 Linkage between PLES Use System and Land Use Types

Following existing studies, this study matches land use types to the PLES use system [31, 32]. The classification is determined by the priority and intensity of functions associated with each land use type. It should be noted that the current land use classification follows the national standard GB/T 21010–2017. The same land use type may be assigned to different space use systems, according to its specific functions, as shown in Fig. 1.3.

As shown in Fig. 1.3, ten sub-categories under cultivated land, grassland, water areas, and urban-rural, industrial, mining, and residential land are classified into the

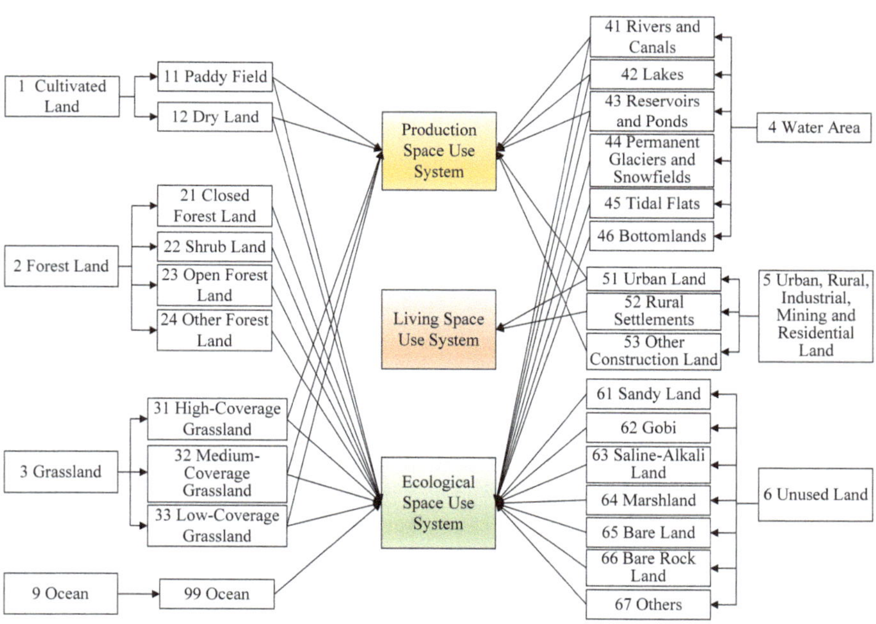

Fig. 1.3 Linkage between production-living-ecological space use system and land use types

production space use system in this study. Specifically, cultivated land is a key type of agricultural production land used for growing various crops. Grassland can serve as pasture for animal husbandry and also supply industrial raw materials. Water areas support multiple functions such as paddy agriculture and fishery production. Urban land and other construction land mainly undertake industrial production functions. The living space use system includes two sub-categories: urban land and rural residential land. They mainly provide places for rest, entertainment, and leisure in human life. The ecological space use system consists of 23 sub-categories under six primary land use types, including cultivated land, forest land, grassland, and water areas. Specifically, crops on cultivated land form part of green infrastructure and constitute important ecological corridors with key ecological functions. Forest land provides ecological services such as climate regulation, water conservation, and biodiversity maintenance. Grassland plays a positive role in atmospheric regulation, soil and water conservation, and ecological landscaping. Water areas perform irreplaceable functions in regulating local climate, conserving water sources, degrading pollutants, and maintaining biodiversity. Unused land has important landscape and ecological functions. In addition, natural land cover types such as sandy land, gobi and saline-alkali land are also important ecological land. Marine land contributes to climate regulation and biodiversity conservation, making it a vital part of the global ecosystem.

1.3 Practical Dilemmas and Research Gaps in Efficiency Evaluation

1.3.1 Theoretical Dilemmas: Multidisciplinary Complexity and Inadequate PLES Coordination

The complexity of multidisciplinary integration and the insufficient coordination of PLES together form the theoretical dilemma in the evaluation of TSUE. These two aspects are interrelated and mutually constraining. Among them, the complexity of multidisciplinary integration is reflected in two key dimensions. On the one hand, there is an inherent difficulty in integrating the geographical perspective of spatial analysis with the economic paradigm of efficiency evaluation. Geography focuses on spatial heterogeneity and spatial correlation, emphasizing the interpretation of dynamic characteristics of spatial use within a geographical context. In contrast, the economic paradigm of efficiency evaluation is centered on input-output theory, emphasizing quantitative measurement and efficiency decomposition. Their theoretical foundations, analytical logic, and evaluation priorities differ substantially. How to achieve an organic integration of the two has become a primary theoretical challenge. This is to avoid evaluation results that either lack rationality in the spatial dimension or deviate from the core goals of efficiency assessment. On the other hand, the quantitative representation and coordinated integration of PLES functions face dual bottlenecks. The three major functions of territorial space are diverse, complex, and interrelated. Some functions, such as ecological regulation and living suitability, are intangible and non-marketized, making them difficult to measure accurately using traditional quantitative methods. Meanwhile, the interactive and restrictive relationships among the three spatial functions are hard to quantify. Furthermore, existing theories have not established a logical framework for the collaborative efficiency evaluation of PLES. As a result, it is difficult for evaluation indicator systems to comprehensively and accurately reflect the comprehensive efficiency of territorial space use. They also fail to support the practical demand for the coordinated optimization of PLES.

1.3.2 Methodological Dilemmas: Bottlenecks in Geospatial Measurement and Prediction Modeling

The core methodological dilemma lies in technical bottlenecks in measuring geospatial effects and predictive modeling. These bottlenecks directly restrict the scientific validity and accuracy of evaluation results and also serve as the core starting point for subsequent model improvement and construction in this study. This dilemma is reflected in two key aspects. Both challenges will be addressed in later research through spatializing traditional evaluation models and integrating multiple techniques in predictive modeling. First, traditional evaluation models have obvious spatial blind spots and struggle to account for both spatial non-stationarity and spatial dependence. There is an urgent need to improve them by incorporating geospatial effects. Currently widely used models, including traditional factor analysis (FA) and data envelopment analysis (DEA) models, assume that evaluation units are independent of each other. They ignore geographical connections in territorial space, fail to capture spatial heterogeneity in space use efficiency across regions, and cannot reflect interactions between regional efficiencies. This leads to inconsistencies between evaluation results and actual spatial conditions. Therefore, integrating geospatial effects into traditional FA and DEA models to realize their spatialization and overcome spatial blind spots has become a major methodological challenge. Second, geospatial effect-oriented predictive modeling suffers from technical limitations, especially insufficient integration between traditional cellular automata (CA) and machine learning techniques. The dynamic evolution of TSUE shows strong spatial correlation and nonlinearity. Traditional rule-driven CA models cannot accurately characterize such complex spatial evolutionary patterns. Although machine learning models have strong nonlinear fitting and feature extraction capabilities, they lack embedded geographical spatial logic. The ineffective integration of the two approaches results in low prediction accuracy of spatial efficiency trends, which cannot provide reliable support for territorial space optimization. In addition, the integration and processing of multi-source heterogeneous spatial data provide fundamental support for the above model improvements and integrations. Remote sensing data, statistical data, geographic vector data, and other data types differ in format, accuracy, and spatiotemporal scale. Problems such as data missing, redundancy, and inconsistent statistical standards are prominent. These issues further increase the technical difficulty of measuring geospatial effects and conducting predictive modeling.

1.3.3 Application Dilemmas: Challenges in PLES Differentiated Evaluation and Targeted Policy Regulation

The core challenges in application are insufficient differentiated evaluation of PLES and weak coordinated policy regulation. These problems hinder the practical application of evaluation results. On the one hand, the lack of differentiated evaluation cannot support targeted policy-making. Most existing studies only measure overall efficiency and analyze spatial distribution. They do not evaluate the three types of spaces separately. Consequently, they fail to identify efficiency shortcomings, evolutionary trends, or key constraints. As a result, evaluation results cannot guide differentiated governance and fail to meet research demands. On the other hand, policy transformation of evaluation results is inefficient. First, efficiency evaluation is poorly linked with territorial spatial planning indicators. Differences in index systems, standards, and objectives make it difficult to integrate evaluation results into planning. Second, dynamic evaluation is temporally mismatched with static planning. The changing spatial efficiency and functional demands conflict with rigid long-term planning. This separates evaluation results from actual planning implementation. Third, coordinated policy regulation is insufficient. There is a lack of targeted policy tools for the three types of spaces. Policies for efficiency improvement lack coordination, making collaborative optimization hard to achieve. This creates a clear gap in policy application and implementation.

1.4 Research Objectives, Core Content, and Scope Definition

1.4.1 Research Objectives

This study takes the modernization of territorial space governance and the coordinated development of PLES as the research background. It aims to construct an integrated evaluation and prediction model system for TSUE by fully considering geospatial effects. By synthesizing multidisciplinary theories and methods including geospatial effects, PLES coordination, and efficiency measurement theories, this study intends to form a complete research chain covering theoretical framework construction, technical method innovation, and policy application transformation. It seeks to provide unified theoretical support and technical tools for promoting the high-quality governance of territorial space.

The specific research objectives are formulated as follows:

1. To systematically review the core theories related to TSUE, geospatial effects and PLES coordination; to clarify the action mechanisms of geospatial heterogeneity, spatial dependence, and spatial spillover in efficiency evaluation; and to lay a solid theoretical foundation and logical basis for subsequent model construction.
2. To construct a geographic factor analysis (GeoFA) model to overcome the defects of traditional FA model in coping with spatial non-stationarity and spatial dependence, so as to realize the accurate screening and optimal selection of evaluation indicators for TSUE.
3. To establish a geographic data envelopment analysis (GeoDEA) model suitable for spatial constraint scenarios, so as to evaluate TSUE and reveal the spatial heterogeneity and interregional interactive dependence of efficiency levels.
4. To develop an extended CA model coupled with machine learning algorithms, so as to simulate the dynamic evolution trend of TSUE and offer forward-looking decision support for territorial space planning.
5. To conduct multi-dimensional evaluation and trend prediction of TSUE taking typical Chinese cities as empirical cases, verify the rationality and applicability of the proposed models, and put forward targeted policy recommendations for territorial space management.

1.4.2 Core Research Content

The core research content of this monograph is divided into the following four parts:

1. **Theoretical Framework Construction**
   This study sorts out core theories, including geospatial effects, input-output theory, production frontier theory, and the PLES theory. It clarifies the connotation and extension of TSUE. This lays the theoretical foundation for subsequent model construction.
2. **Model Construction and Optimization**
   Based on the traditional FA model, geospatial effects are introduced to build the GeoFA model. This resolves the problem of spatial non-stationarity in indicator screening. Built on the classic DEA model, spatial constraints are integrated to establish the GeoDEA model. This achieves accurate evaluation of spatial efficiency. Machine learning algorithms, including convolutional neural network (CNN), long short-term memory (LSTM), and random forest (RF) models, are combined with the traditional CA model to construct an extended CA model. This model simulates the dynamic changes of TSUE.

3. **Empirical Research**
   Taking typical Chinese cities as the study area, an evaluation index system for PLES use efficiency is constructed. The GeoFA, GeoDEA, and extended CA models are applied to conduct multi-dimensional evaluation and prediction. This study reveals the spatiotemporal evolution characteristics and spatial autocorrelation of TSUE.
4. **Policy Recommendations**
   Based on empirical results, this study proposes dimension-specific and zoned policy frameworks and pathways for improving TSUE. It provides decision support for territorial space planning and management.

1.4.2.1 Scope Definition

1. **Spatial Boundary**
   This study takes 340 prefecture-level cities (prefectures and leagues) across China as basic evaluation units. It focuses on spatial disparities, spatiotemporal evolution, and future trends of TSUE at the municipal and regional scales.
2. **Temporal Boundary**
   The empirical study covers the period from 1990 to 2030. Historical data for 1990, 2000, 2010, and 2020 are used to analyze spatiotemporal evolution characteristics. Data for 2030 are predicted to simulate future changing trends of TSUE.
3. **Content Boundary**
   This study concentrates on the evaluation and prediction of TSUE. It does not involve the specific formulation or implementation of territorial space planning. Nevertheless, it provides a scientific basis and decision support for territorial space planning.

1.5 Technical Roadmap

As shown in Fig. 1.4, the study structure and chapter logical framework are presented.

Stage 1: Theoretical Analysis (Chaps. 1–2)—Theoretical Foundation and Identification of Research Gaps

This stage lays a solid theoretical foundation for the entire research. It systematically reviews the historical context and strategic significance of TSUE evaluation, defines core concepts, and elaborates the relevant theoretical basis. Existing evaluation approaches are carefully assessed to identify practical dilemmas and research gaps in theoretical integration, methodological modeling, and practical implementation. The evolutionary trajectory of models associated with TSUE is traced, and

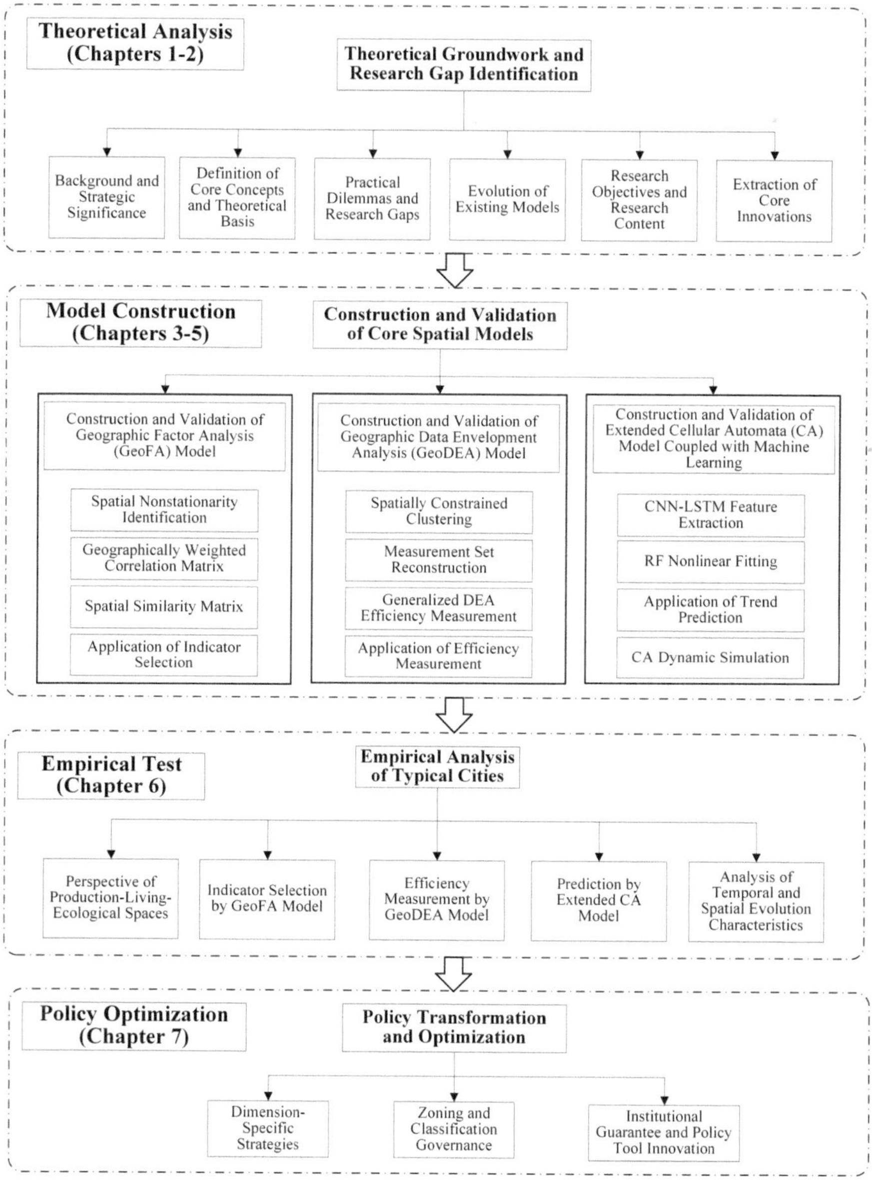

Fig. 1.4 Logical framework and organizational structure of the monograph

the strengths and spatial limitations of traditional models are summarized. Based on the above analysis, this study clarifies its research objectives, core contents, and key innovations, offering clear guidance for the subsequent development of spatialized models and empirical investigations.

Stage 2: Model Construction (Chaps. 3–5)—Construction and Validation of Core Spatialized Models

Responding to the demand for explicit representation of geospatial effects, this stage sequentially constructs and validates three core spatialized models, forming an integrated technical framework that overcomes the spatial expression limitations of traditional models.

The GeoFA model is established to identify spatial non-stationarity characteristics. Through dual-path modeling using a geographically weighted correlation matrix and a spatial similarity matrix, this model realizes the spatialized improvement and application of indicator screening methods, supporting the construction of an efficiency evaluation indicator system for PLES.

The GeoDEA model is constructed to reconstruct the measurement set via spatially constrained clustering. Integrated with the generalized DEA method, it optimizes efficiency measurement in a spatialized manner and achieves accurate measurement and empirical application of TSUE.

An extended CA model coupled with machine learning is developed. Spatial-temporal feature extraction is implemented using the CNN-LSTM model, while complex nonlinear relationships are fitted via RF. Combined with the CA dynamic simulation engine, a trend prediction model integrating data-driven and rule-constrained mechanisms is established, enabling high-precision prediction and application validation of TSUE.

Stage 3: Empirical Examination (Chap. 6)—Empirical Analysis of Typical Chinese Cities

Taking typical cities in China as study areas, this stage conducts a full-process empirical test from the PLES perspective to verify the scientificity and practicability of the proposed model system. From the core perspective of PLES, the GeoFA model is used to screen dimension-specific efficiency evaluation indicators for production, living, and ecological spaces. The GeoDEA model is applied to accurately measure the dimensional and overall TSUE, while the extended CA model is employed to predict the trend of TSUE by 2030. Finally, the spatiotemporal evolution characteristics and spatial correlation patterns of TSUE from 1990 to 2020 are systematically analyzed, and the overall features and evolutionary trends of urban TSUE in China are summarized.

Stage 4: Policy Optimization (Chap. 7)—Policy Transformation and Optimal Design

Based on empirical results, this stage realizes the logical connection from efficiency measurement analysis to spatial governance practice, promoting the translation of research findings into practical applications. Differentiated efficiency improvement strategies are formulated for PLES, forming a zoning and classification governance framework consistent with the spatial pattern of efficiency. Meanwhile, the institutional guarantee system is improved, and policy tools are innovated, facilitating the deep integration of efficiency evaluation results into the whole process of territorial spatial planning, including formulation, implementation, and supervision. Ultimately, this stage provides implementable policy support for the coordinated optimization of PLES, high-quality development of territorial space, and the modernization of urban spatial governance.

1.6 Core Innovations

The innovations of this study are manifested in three interconnected dimensions.

First, methodological innovation. Traditional models typically assume global homogeneity. This study challenges that assumption by introducing geospatial effects. An integrated framework combining geographical perspectives with classical models is established. This framework upgrades indicator selection, efficiency measurement, and trend prediction into spatially explicit processes. Thus, the technical bottleneck of traditional models, namely their failure to capture spatial heterogeneity and dependence, is effectively addressed.

Second, model system innovation. A full-chain spatial modeling system is constructed. It covers three stages: indicator screening, efficiency measurement, and trend prediction. The GeoFA, GeoDEA, and extended CA models are organically integrated within this system. The result is a complete methodological framework for evaluating TSUE. Both the scientific rigor and the precision of evaluation results are significantly enhanced.

Third, practical application innovation. This study responds to two pressing demands: China's territorial spatial planning reform and ecological civilization construction. Typical cities serve as empirical cases. Model applications are actively linked with policy formulation. Dimension-specific, zoned, and classified pathways for efficiency improvement are proposed. These pathways bridge the gap between evaluation outcomes and policy practice. Technical support and practical references are thus provided for modernizing territorial spatial governance.

Competing Interests The authors have no conflicts of interest to declare that are relevant to the content of this chapter

References

1. Xie X, Fang B, Xu H.Z.Y et al (2021) Study on the coordinated relationship between urban land use efficiency and ecosystem health in China. Land Use Policy 102:105235
2. Xue D, Yue L, Ahmad F et al (2021) Urban eco-efficiency and its influencing factors in western China: fresh evidence from Chinese cities based on the US-SBM. Ecol Indic 127:107784

3. Gao JX, Song JB, Wu LF (2022) A new methodology to measure the urban construction land-use efficiency based on the two-stage DEA model. Land Use Policy 112:105799
4. Pu WF, Zhang AL (2022) Can China's market-oriented reform improve the efficiency of industrial land use? A panel data empirical analysis at prefecture level from 2007-2019. Front Environ Sci 10:884958
5. Ma Y, Zheng MR, Zheng XQ et al (2023) Land use efficiency assessment under sustainable development goals: a systematic review. Land 12(4):894
6. Chen YF, He CK, Guo W et al (2023) Mapping urban functional areas using multisource remote sensing images and open big data. IEEE Journal of Selected Topics in Applied Earth Observations and Remote Sensing 16:7919–7931
7. Li Q, Cui CH, Liu F et al (2022) Multidimensional urban vitality on streets: spatial patterns and influence factor identification using multisource urban data. ISPRS Int J Geo Inf 11(1):2
8. Li HY, Hu MM, Li G (2020) Forecasting tourism demand with multisource big data. Ann Tour Res 83:102912
9. Fang XY, Jiang WM, Miao SG et al (2004) The multi-scale numerical modeling system for research on the relationship between urban planning and meteorological environment. Adv Atmos Sci 21(1):103–112
10. Samuelson PA, Nordhaus WD (2002) Microeconomics. Huaxia Publishing House, Beijing
11. Sun YF, Jia Z, Chen Q et al (2023) Spatial pattern and spillover effects of the urban land green use efficiency for the Lanzhou-Xining urban agglomeration of the yellow river basin. Land 12(1):59
12. Wang YB, Zhao RT, Li Y et al (2023) Research on the evolution and the driving forces of land use classification for production, living, and ecological space in China's Qilian Mountains nature reserve from 2000 to 2020. Environ Sci Pollut Res 30(24):64949–64970
13. Zhao J, Zhao YL (2023) Synergy/trade-offs and differential optimization of production, living, and ecological functions in the Yangtze River economic belt. China Ecological Indicators 147:109925
14. Liu Y, Wang KL, Xing XY et al (2023) On spatial effects in geographical analysis. Acta Geograph Sin 78(03):517–531
15. Zhao JY, Simpson M, Wallgrun JO et al (2020) Exploring the effects of geographic scale on spatial learning. Cognitive Research-Principles and Implications 5(1):14
16. Dietzenbacher E, Lenzen M, Los B et al (2013) Input-output analysis: the next 25 years. Econ Syst Res 25(4):369–389
17. Chen WC, McGinnis LF (2007) Reconciling ratio analysis and DEA as performance assessment tools. Eur J Oper Res 178(1):277–291
18. Aldieri L, Gatto A, Vinci CP (2021) Evaluation of energy resilience and adaptation policies: an energy efficiency analysis. Energy Policy 157:112505
19. Dyckhoff H, Souren R (2022) Integrating multiple criteria decision analysis and production theory for performance evaluation: framework and review. Eur J Oper Res 297(3):795–816
20. Peykani P, Memar-Masjed E, Arabjazi N et al (2022) Dynamic performance assessment of hospitals by applying credibility-based fuzzy window data envelopment analysis. Healthcare 10(5):876
21. Li W, Xue CX, Zhu RX et al (2012) Analysis on production allocative efficiency of agricultural machinery based on frontier theory in China. Transactions of the CSAE 28(03):38–43
22. Ferreira DC, Figueira JR, Greco S et al (2023) Data envelopment analysis models with imperfect knowledge of input and output values: an application to portuguese public hospitals. Expert Syst Appl 231:120543
23. Zhou ZX (2014) Research on theory, modelling and application of integer DEA. Dissertation,. University of Science and Technology of China
24. Hou YZ, Zhang ZL, Wang YR et al (1954) (2022) function evaluation and coordination analysis of production-living-ecological space based on the perspective of type-intensity-connection: a case study of Suzhou, China. Land 11(11)
25. Nguyen TT, Verdoodt A, Tran VY et al (2015) Design of a GIS and multi-criteria based land evaluation procedure for sustainable land-use planning at the regional level. Agric Ecosyst Environ 200:1–11

26. Wang Q, Wang HJ (2022) Dynamic simulation and conflict identification analysis of production-living-ecological space in Wuhan, Central China. Integr Environ Assess Manag 18(6):1578–1596
27. Duan YM, Wang H, Huang A et al (2021) Identification and spatial-temporal evolution of rural "production-living-ecological" space from the perspective of villagers' behavior-a case study of Ertai town. Zhangjiakou City Land Use Policy 106:105457
28. Fu C, Tu XQ, Huang A (2021) Identification and characterization of production-living-ecological space in a central urban area based on POI data: a case study for Wuhan. China Sustainability 13(14):7691
29. Fu JY, Bu ZQ, Jiang D et al (2022) Sustainable land use diagnosis based on the perspective of production-living-ecological spaces in China. Land Use Policy 122:106386
30. Yang YY, Bao WK, Liu YS (2020) Coupling coordination analysis of rural production-living-ecological space in the Beijing-Tianjin-Hebei region. Ecol Indic 117:106512
31. Liao GT, He P, Gao XS et al (2022) Land use optimization of rural production-living-ecological space at different scales based on the BP-ANN and CLUE-S models. Ecol Indic 137:108710
32. Liu JL, Liu YS, Li YR (2017) Classification evaluation and spatial-temporal analysis of "production-living-ecological" spaces in China. Acta Geograph Sin 72(07):1290–1304

Chapter 2
Evolution of Models for Territorial Space Use Efficiency Research

Abstract This chapter focuses on the evolution of models for TSUE research. It first traces the origin and development of TSUE, which derives from land use efficiency, and clarifies its close connection with national policies, especially the shift from "land use" to "territorial space use" in China. Then, it systematically reviews the research on indicator system construction models, noting their evolution from single-factor to multi-factor and ultimately to comprehensive frameworks, while pointing out the subjectivity in indicator selection and neglect of geospatial effects. It also classifies efficiency measurement models into three types and analyzes their advantages and limitations, emphasizing that traditional models fail to consider geospatial effects. Finally, it summarizes the application of prediction models and their shortcomings, concluding that current research lacks sufficient consideration of geospatial effects, laying a foundation for subsequent model improvement.

2.1 Origin and Evolution of Territorial Space Use Efficiency

The concept of TSUE is gradually derived from that of land use efficiency [1]. As an early form and foundational stage of its development, land use efficiency has a long research history and a relatively mature theoretical system across the global academic community. An in-depth exploration of its origin provides valuable theoretical support and practical implications for this study. Early Marxist views of nature emphasized the coexistence and harmonious relationship between humans and nature at an appropriate scale, laying a philosophical foundation for the efficient and sustainable use of land resources [2]. Scholars from morphology, urban planning, geography, regional economics, and other disciplines subsequently conducted extensive research on the spatial distribution and evolutionary characteristics of land use [3, 4]. This led to the proposal of several influential land use models, such as the concentric zone model, sector model, polycentric model, and development axis model. A series of classical land use theories also took shape, including von Thünen's agricultural location theory, land location theory, the law of diminishing marginal

Y. Ma, *Evaluation and Prediction Modeling of Territorial Space Use Efficiency*, Urban Sustainability, https://doi.org/10.1007/978-981-92-4396-9_2

returns, smart growth, and human–land relationship theory [5]. The core purpose of these models and theories is to maximize the output benefits of land use and improve land use efficiency through policy and technical instruments that fully tap the developmental potential of land. On this basis, international scholars have carried out empirical and case-based research to examine economic efficiency, social efficiency, and ecological efficiency in land use activities [1, 6]. In the early nineteenth century, for instance, Germany and the United States launched research projects on land use, using case analyses to inform the formulation of policies, regulations, and standards for efficient land utilization. Since the twentieth century, overseas researchers have further explored the efficient and sustainable development of land use from multiple perspectives. These include the relationship between agricultural production efficiency and ecological conditions, the linkage between urban growth and land use efficiency, and the evaluation of urban land use efficiency [7–10].

Certain countries initiated research on land use efficiency at a relatively late stage. China is a typical case in this regard. Land use assessments mentioned in taxation practices during the mid-twentieth century can be regarded as the preliminary exploration of land use efficiency research in China [11]. Following the reform and opening-up, rapid economic expansion in China triggered a series of land-related problems, including the unregulated sprawl of construction land, extensive utilization of land resources, and declining output efficiency per unit land area [12, 13]. In response to these challenges threatening the sustainable development of land use, land use evaluation and efficiency measurement have become key research topics in geography, economics, sociology, and other disciplines [12]. For example, Yu et al. applied principal component analysis and the FA model to explore the causes of cultivated land loss and estimate cultivated land use efficiency in Jiangyin City, representing one of the earliest quantitative studies on land use efficiency in China [14]. In recent years, scholars have extended their research to multiple dimensions, including urban land use efficiency, industrial land use efficiency, and ecological efficiency of land use [11–13]. These studies have contributed significantly to theoretical progress while providing strong support for addressing practical land use issues. Meanwhile, research progress and trends in China's land use efficiency have been closely tied to national policies due to the country's specific institutional context. Accordingly, this monograph reviews the evolution of relevant policies to establish a research paradigm that better aligns with real-world applications.

Chinese authorities have attached great importance to land use efficiency. In 2008, the State Council issued the Circular on Promoting Economical and Intensive Land Use. This document emphasized the principle of economical and intensive land use, with the clear goal of improving land use efficiency. Its release marked the beginning of national-level attention and regulation toward efficient land utilization. Subsequently, the government introduced a series of policy documents. These policies focused on the protection, optimization, and efficiency improvement of land use patterns. In 2019, the Central Committee for Comprehensive Deepening of Reform issued the Several Opinions on Establishing a Territorial Spatial Planning System and Supervising Its Implementation. This document formally integrated multiple spatial plans into a unified territorial spatial plan. It signaled a conceptual shift from "land use" to "territorial space use". This shift also encouraged scholars to extend their research from land use to territorial space use, promoting deeper exploration in this field. In addition, the document set a long-term goal to achieve by 2030. By that time, production space is expected to be intensive and efficient, living space pleasant and appropriate, and ecological space clean and attractive. To achieve this goal, accurate evaluation and analysis of TSUE are essential. Such evaluation helps identify existing problems and supports the formulation of effective policy responses.

2.2 Research on Indicator System Design Models

Constructing an index system for the quantitative measurement of TSUE has been a widely adopted research approach among scholars. The rationality and scientific validity of the index system directly determine the accuracy of efficiency measurement. Owing to differences in national development stages and land use systems, scholars from various countries have developed distinct index systems for quantifying TSUE.

In the early stage, researchers preferred to build index systems from a single-factor perspective to measure the use efficiency of different land types, which refers to the total output benefit per unit land area [15]. For example, Hernandez et al. quantified the land use efficiency of solar energy land by calculating the benefits produced by solar installations per unit area [16]. Zitti et al. adopted per capita built-up area to represent urban land use efficiency and explored the relationship between land growth and land use efficiency [17]. Melchiorri et al. measured residential land use efficiency by calculating the ratio of land consumption rate to population growth rate under the background of sustainable development goals [18]. In recent years, with the deepening understanding of territorial space, foreign researchers have gradually established more complex multi-factor index systems to measure efficiency [19]. For instance, scholars from the United States, France, the United Kingdom, and other countries have constructed an index system from four aspects to evaluate the use efficiency of construction land. These aspects include building height, building density, floor area ratio, and the proportion of open space. In addition, an in-depth review of foreign literature shows that researchers mainly select indicators based on land use type, disciplinary background, previous research experience, data availability, and other factors. Few studies use modeling methods to select indicators.

With the continuous evolution of national land policies and development strategies, the selection of indicators for measuring TSUE has evolved from single factors to multiple factors and then to comprehensive factors. Initially, scholars adopted a single input-output ratio to measure efficiency values. For example, Li et al. used the added value of secondary and tertiary industries per unit urban built-up area to evaluate urban land use efficiency [20]. Liu et al. took per-mu yield or per-mu profit as indicators to study agricultural land use efficiency and analyzed the relationship between agricultural land use efficiency and farm household operation scale [21].

Although single-factor indicators have been widely used due to their simple calculation, territorial space use is a complex system. The output benefit per unit land area cannot fully and accurately reflect TSUE, which may lead to errors in efficiency measurement. Therefore, a growing number of scholars have begun to construct indicator systems for measuring TSUE based on multi-factor input-output theory [22]. For instance, Tian et al. selected 12 indicators from six perspectives, including land, capital, and expected output [23]. These indicators include construction land area, fixed assets, sulfur dioxide emissions, and other items. They established an indicator system for measuring industrial land use efficiency in the Beijing-Tianjin-Hebei region and analyzed the spatial heterogeneity of industrial space in the Beijing-Tianjin-Hebei urban agglomeration. Ma et al. selected measurement indicators of land use ecological efficiency from an input-output perspective [22]. The indicator set includes energy, labor, and capital as inputs and economic output, green coverage rate, and pollutants as outputs. They further explored the coupling and coordination relationship between land use ecological efficiency and green infrastructure. Since the introduction of the multiple-planning integration and one-map policies, the academic community has paid more attention to the overall TSUE. Researchers mainly conduct comprehensive measurements by classifying territorial space types. As a result, efficiency measurement indicator systems have become more comprehensive and integrated. For example, Wang et al. analyzed territorial space from the perspective of production, living, and ecological spaces [24]. They constructed a 27-indicator system for measuring efficiency that covers the use of production, living, and ecological spaces. Yang et al. established an input-output indicator system based on the functional orientation of territorial space [25]. The system includes territorial development space, agricultural production space, ecological environment space, and sustainable support space. It is used to evaluate the development and utilization efficiency of territorial space at the county level.

The indicator system for measuring TSUE is gradually developing toward multi-level, multi-space, and multi-indicator dimensions. This evolution is mainly driven by two factors. First, with the deepening understanding of territorial space use and the continuous advancement of research on TSUE, researchers have increasingly recognized that multi-level and diversified indicator systems can measure efficiency more comprehensively and accurately. Second, with the rapid development of surveying and geographic information technology, data acquisition channels have been expanding continuously. The emergence of multi-source data has provided richer and more comprehensive data support for research on TSUE. This has further promoted the evolution of indicator systems. However, such evolution has posed certain challenges to the rational selection of indicators. At present, researchers mostly select indicators by referring to existing research results. This practice leads to subjectivity and irrationality in indicator selection. In addition, although multi-source data have broadened data acquisition channels and enriched the types of indicator data, they have also increased the difficulty of indicator selection and determination. When facing massive, complex, and multi-dimensional data, employing models to sort out data and reduce dimensionality is recognized as an effective research approach. Although some scholars have assisted indicator selection by adopting methods such as the analytic hierarchy process and the FA model, most of the models applied are simple mathematical models [22]. Territorial space is a kind of geographic space with coordinate location, geographical position, and social-economic attributes. The geographical spatial effect is particularly significant among different territorial spaces. Therefore, when constructing an indicator system for measuring TSUE, it is necessary to fully recognize the impact of geospatial effects on various geographic spatial data and social-economic data carried by territorial space. Nevertheless, some scholars often neglect geospatial effects when using simple mathematical models to reduce dimensions and select indicators for TSUE. This negligence consequently affects the rationality and accuracy of indicator system construction.

In summary, the indicator system for measuring TSUE is developing toward integration with multiple levels, multiple spatial dimensions, and multiple indicators. The application of modeling methods for indicator selection from multi-source data offers a feasible approach to establishing a scientific and rational indicator system. However, this approach requires more reliable models as technical support. Such models should fully consider the influence of geospatial effects on indicator selection during operation. This practice can help ensure that the constructed indicator system is more reasonable and scientific.

2.3 Research on Efficiency Measurement Models

Establishing a reasonable indicator system is a prerequisite for measuring TSUE. Selecting a scientific efficiency measurement model lies at the core of the entire measurement process. A comprehensive review of extensive literature shows that studies on efficiency measurement models by researchers worldwide are largely consistent. With the continuous enrichment and deepening of research on land use and territorial space, models for measuring TSUE have gradually evolved and upgraded. Measurement models have gradually evolved from qualitative descriptive models to quantitative data-driven models. A relatively complete system of efficiency measurement models has been established in the academic community. This study classifies measurement models into three categories based on differences in indicator systems. They are single-indicator measurement models, multi-indicator comprehensive measurement models, and input-output indicator measurement models, as shown in → Fig. 2.1.

First, single-indicator measurement models represent the most basic, simplest, and most applicable models in the measurement of TSUE. The United Nations Sustainable Development Goal 11.3.1 measures land use efficiency by calculating the ratio of land consumption rate to population growth rate [26, 27]. Kilkis defines land use efficiency by calculating the ratio of construction land to population and examines the relationship between land use efficiency and urban emissions in different regions [28]. Chakraborty regards urban density as an effective indicator for measuring urban land use efficiency [29]. This scholar measures land use efficiency by calculating the

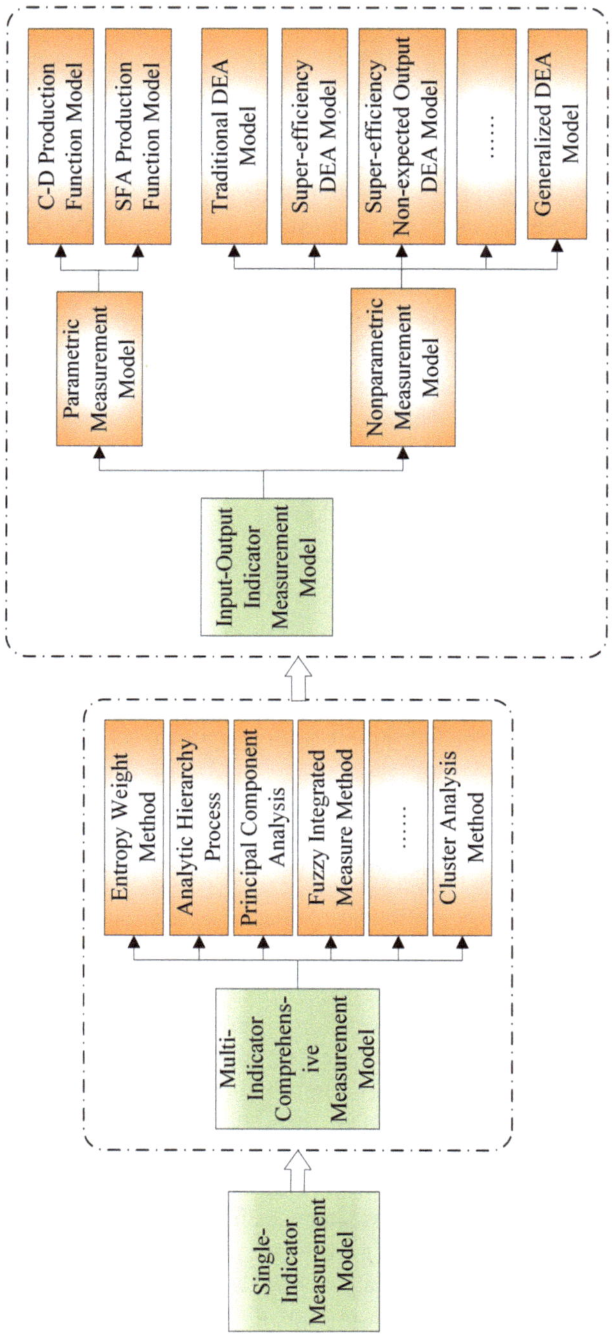

Fig. 2.1 The evolution of measurement models for territorial space use efficiency

ratio of urban population to built-up area and explores the relationship between urban expansion patterns and land use efficiency at the global scale. Li et al. measure urban land use efficiency by calculating the added value of secondary and tertiary industries per unit built-up area [20]. Ma et al. define the ratio of population to construction land as land use efficiency from a demographic perspective. They also define the ratio of gross domestic product to construction land as land use efficiency from an economic perspective. On this basis, the authors analyze the coordinated relationship of land use efficiency under the two perspectives [30].

Second, multi-indicator comprehensive measurement models have attracted extensive attention from researchers. The core of such models lies in determining the weights of multiple indicators. TSUE is then measured through weighted summation. These models are mainly based on mathematical theories. They feature strong comprehensiveness, wide applicability, and high comparability. Commonly used comprehensive measurement models include the entropy weight method, analytic hierarchy process, principal component analysis, fuzzy data comprehensive evaluation method, cluster analysis, discriminant analysis, Bi-TOPSIS method, matter-element model, and others [15]. Langpap et al. combined econometric models and comprehensive measurement models to explore land use efficiency [31]. They also analyzed the impacts of land use policies on land use efficiency. Makowski et al. comprehensively measured agricultural land use efficiency by improving the linear programming model [32]. In China, many researchers have adopted comprehensive measurement models to calculate efficiency values. For example, Cao et al. applied cluster analysis and discriminant analysis to measure and analyze urban construction land use efficiency [33]. Zhang et al. conducted an empirical study on land use efficiency in 13 cities of Jiangsu Province using the Bi-TOPSIS model [34]. They further discussed the convergence characteristics of urban land use efficiency. Although comprehensive measurement models are favored by some researchers, they still have certain limitations in practical application. Specifically, these models lack support from economic theories. They also suffer from difficulties in connotation definition, marked subjectivity in indicator weight assignment, and ambiguity of ideal values. These problems have affected the objectivity and accuracy of efficiency measurement results to a certain extent [4].

Finally, measurement models based on input-output indicators are constructed on the basis of economics, operations research, and mathematical theories. These models are the most widely applied, diverse, and highly recognized among efficiency measurement approaches. They are mainly divided into two categories, namely parametric measurement models and non-parametric measurement models, as shown in → Fig. 2.1.

Parametric measurement models mainly include the Cobb-Douglas production function and stochastic frontier analysis (SFA) [35, 36]. Scholars apply such models to measure land use efficiency of different types. For example, Ferreira et al. constructed an SFA model based on agricultural census data [6]. They quantitatively measured land use efficiency in the Brazilian Amazon and examined the relationship between farm size and land use efficiency. Jin et al. adopted the Cobb-Douglas production function and input-output panel data [37]. They measured land use efficiency in 110 cities along the Yangtze River Economic Belt and explored their spatial correlation characteristics. Ding et al. used the SFA model to calculate land use efficiency and ecological efficiency of urban agglomerations in the Yellow River Basin [38]. They also conducted spatial autocorrelation analysis. The core idea of parametric measurement models is to introduce land as a production factor into the production function to achieve efficiency measurement. Such models take into account the influence of random factors on efficiency measurement. However, in practical application, the production function needs to be specified in advance, which leads to strong subjectivity. Meanwhile, this type of model is not suitable for studies with multiple output requirements.

Non-parametric measurement models are the most widely applied models in the measurement of TSUE. According to statistics, studies employing such models account for approximately 45% of the relevant literature [15]. These models do not require the establishment of a production function for parameter estimation. They avoid random errors caused by parameter estimation and thus achieve more objective efficiency measurement. In addition, such models are capable of handling multiple input and output issues [39]. As a typical representative of non-parametric measurement models, the DEA model is constructed on the basis of operations research and economic theories. After more than 40 years of development and evolution, it has formed a large family of DEA models [40, 41]. Initially, the two most basic models were the CCR model under constant returns to scale and the BCC model under variable returns to scale. Subsequently, researchers took strong efficiency, weak efficiency, and slack variables into consideration. They discussed distance functions to the frontier, including the farthest distance to the frontier, also known as the Slack-Based Measure (SBM)model, the shortest distance to the strong efficient frontier, the directional distance function, the shortest distance to the weak efficient frontier, and the hybrid distance function [42, 43]. Finally, a series of extended models have gradually been derived. These include the super-efficiency model, the super-efficiency model with undesirable outputs, the generalized DEA model, and the Malmquist model [44, 45]. DEA series models have been widely favored by researchers. For example, Pascual et al. used the DEA model to measure agricultural land use efficiency and attempted to enhance the intensive potential of land use by improving technical efficiency [10]. Lu et al. applied the super-efficiency SBM model to measure urban land use efficiency in national high-tech zones in China from 2003 to 2016 [46]. They further explored the impact of the pilot free trade zone policy on urban land use efficiency.

Based on the above summary and discussion, most existing measurement models of TSUE are constructed within theoretical frameworks of economics, operations research, and mathematics. These models were originally designed to measure the efficiency of relatively independent DMUs. Typical examples include corporations, banks, and hospitals. Such models do not consider territorial spaces as DMUs with geospatial effects. As a basic discipline studying human-land relationships, geography provides important theoretical support for research on TSUE [47, 48]. According to fundamental principles of geography, geospatial effects serve as a key factor influencing measurement results of TSUE [15]. However, direct application of these traditional models to measuring the efficiency of territorial spatial units often leads to problems. These problems include an unreasonable reference frontier and inaccurate efficiency measurement. Therefore, to obtain more reliable efficiency measurement results, the measurement models of TSUE need to be reconstructed and improved.

2.4 Research on Efficiency Prediction Models

In an era of rapid advances in computer technology, a large number of prediction models have been developed. These models enable future projections based on existing time-series data and spatial evolution characteristics. Wide applications of such models can be found in geography, sociology, economics, medicine, and other disciplines. This approach offers researchers and policymakers a basis for anticipating future development trends [49]. Research on TSUE addresses a fundamental land issue that has long drawn key national attention. This field also forms an important part of territorial space monitoring and early warning systems. Future trends in this domain deserve high expectations and considerable attention.

A review of extensive literature reveals that various models have been widely applied by researchers to the simulation and prediction of land use. These models include CA, Markov Model, System Dynamics Model, RF, CNN, and LSTM [50, 51]. Applications of these models have laid a solid foundation for the prediction of TSUE. These models work in a coordinated and complementary manner. Strong technical support for prediction research is formed through mutual reinforcement among different model frameworks. For instance, some researchers have coupled the CA model with the Markov model [52, 53]. This combination supports accurate prediction of land use and land cover under different scenarios. The integrated model contributes to the sustainable development of ecological environments and future megacities. Other researchers have integrated machine learning models with the CA model [53–55]. Full capture of spatiotemporal features within such frameworks greatly improves the accuracy of land use and land cover prediction. In addition, continued progress in interdisciplinary research has promoted the coupling of models from different fields. Such models include the ABM model, GIS model, and SD model [56]. This integration has brought new vitality to research on land use and land cover prediction. Research on land use simulation has attracted extensive attention from scholars. Continuous empirical attempts have led to the gradual maturity of relevant prediction models. This progress provides a solid basis for studies on the prediction of TSUE.

Research on TSUE started later than studies on general land use. Future development trends of TSUE have still aroused extensive concern among researchers. In recent years, scholars have begun to employ prediction models to forecast efficiency values. For example, Verburg et al. measured European land use efficiency based on comprehensive evaluation methods and the Global Trade Analysis Project [57]. A series of landscape-level simulation models were adopted to realize dynamic simulation of land use. Lu et al. applied the CA and Markov models to simulate future urban land use changes in Tianjin under environmental protection scenarios [58]. This work indirectly predicted future urban land use efficiency and provided key data and information for land use planning. Liang adopted the grey GM (1,1) model, which allows modeling with small samples, to predict construction land use efficiency [59]. Corresponding countermeasures for improving construction land use efficiency were proposed according to prediction results. Hao used multiple methods including Markov state transition matrix, grey dynamic model GM (1,1) and Vector Auto-Regressive model [60]. These methods were applied to predict ecological efficiency values, spatial distribution, total factor productivity, and its decomposition indicators at the provincial level in China. Related results provide data support and references for the coordinated development of economic and ecological efficiency. Based on the above analysis, explorations on prediction models for TSUE remain limited compared with studies on land use prediction models. Among existing research efforts, the grey dynamic model GM (1,1) is widely adopted for its ability to conduct prediction with small sample sizes. This model is not suitable for long-term sequence prediction. It also fails to fully consider the impacts of spatial factors on prediction results. In practice, the evolution of TSUE is a long-term and complex process. The evolution is affected not only by long-time series but also by neighboring spatial conditions. Therefore, a reliable prediction model with strong capability in capturing and expressing spatiotemporal features is necessary. Such a model can support a more accurate grasp of future development trends of TSUE.

2.5 Chapter Summary

This chapter systematically reviews research progress on the origin and evolution of TSUE, as well as corresponding measurement indicator systems, efficiency measurement models, and prediction models. A relatively complete theoretical and methodological system has been formed in existing studies. Indicator systems have gradually developed from single-factor frameworks to multi-factor comprehensive frameworks. Efficiency measurement models fall into three categories: single-indicator models, multi-indicator comprehensive models and input-output models. Prediction models applied in land use research also provide references for trend analysis of efficiency. However, insufficient consideration of geospatial effects is common in current research on indicator screening, efficiency measurement, and trend prediction. Problems include strong subjectivity in indicator selection, ignorance of spatial heterogeneity in traditional measurement models, and weak spatiotemporal representation ability of prediction models. Further improvements are still needed in relevant research.

Competing Interests The authors have no conflicts of interest to declare that are relevant to the content of this chapter.

References

1. Hu PR (2023) Study on evaluation of efficiency and strategy of optimization in territory spatial utilization of Taikang county. Dissertation, Huazhong University of Science and Technology
2. Li J (2022) Research on the temporal and spatial patterns and influencing factors of the comprehensive efficiency of land and space utilization in cities. Dissertation, China University of Geosciences (Beijing)
3. Niu B, Ge DZ, Sun JW et al (2023) Multi-scales urban-rural integrated development and land-use transition: the story of China. Habitat Int 132:102744
4. Yang B (2021) Study on land space use efficiency in the context of ecological civilization: a case study of the urban agglomeration in the middle reaches of the Yangtze River, China. Dissertation, China University of Geosciences (Wuhan)
5. Liu SH, Wu CJ, Chen T (2001) A critical review on the progress of urban land use theories in the West. Geogr Res 1:111–119
6. Ferreira MDP, Féres JG (2020) Farm size and land use efficiency in the Brazilian Amazon. Land Use Policy 99:104901
7. Barbosa JA, Bragança L, Mateus R (2015) Assessment of land use efficiency using bsa tools: development of a new index. J Urban Plan Dev 141(2):04014020
8. Koroso NH, Lengoiboni M, Zevenbergen JA (2021) Urbanization and urban land use efficiency: evidence from regional and Addis Ababa satellite cities, Ethiopia. Habitat Int 117:102437
9. Koroso NH, Zevenbergen JA, Lengoiboni M (2020) Urban land use efficiency in Ethiopia: an assessment of urban land use sustainability in Addis Ababa. Land Use Policy 99:105081
10. Pascual U (2005) Land use intensification potential in slash-and-burn farming through improvements in technical efficiency. Ecol Indic 52(4):497–511
11. Xue D, Yue L, Ahmad F et al (2021) Urban eco-efficiency and its influencing factors in western China: fresh evidence from Chinese cities based on the US-SBM. Ecol Indic 127:107784
12. Jiang HL (2021) Spatial-temporal differences of industrial land use efficiency and its influencing factors for China's central region: analyzed by SBM model. Environ Technol & Innov 22:101489
13. Ouyang X, Wei X, Wei GE et al (2023) The expansion efficiency of urban land in China's urban agglomerations and its impact on ecosystem services. Habitat Int 141:102944
14. Yu YJ, Lu YL (2002) Quantitative studies on the using efficiency and driving factors of changing of cultivated land in Jiangyin City. Econ Geogr 4:440–443+447
15. Ma Y, Zheng MR, Zheng XQ et al (2023) Land use efficiency assessment under sustainable development goals: a systematic review. Land 12(4):894
16. Hernandez R R, Hoffacker M K. Field C B (2014) Land-use efficiency of big solar. Environ Sci & Technol 48(2):1315-1323
17. Zitti M, Ferrara C, Perini L et al (2015) Long-term urban growth and land use efficiency in southern Europe: implications for sustainable land management. Sustainability 7(3):3359–3385
18. Melchiorri M, Pesaresi M, Florczyk AJ et al (2019) Principles and applications of the global human settlement layer as baseline for the land use efficiency indicator-DSDG 11.3.1. ISPRS Int J Geo-Inf 8(2):96
19. Schiavina M, Melchiorri M, Freire S et al (2022) Land use efficiency of functional urban areas: Global pattern and evolution of development trajectories. Habitat Int 123:102543
20. Li YL, Shu BR, Wu Q (2014) Urban land use efficiency in China: spatial and temporal characteristics, regional difference and influence factors. Econ Geogr 34(01):133–139
21. Liu Y (1992) Liu X J (2025) The impact of "land and services" dual-scale management on agricultural operational benefit: A comparison with land-scale management. Land 14(10)
22. Ma Y, Zheng MR, Xu F et al (2023) Modeling and exploring the coordination relationship between green infrastructure and land use eco-efficiency: an urban agglomeration perspective. Environ Sci Pollut Res 30(40):92537–92554
23. Tian YY, Zhou DY, Jiang GH (2021) A new quality management system of admittance indicators to improve industrial land use efficiency in the Beijing-Tianjin-Hebei region. Land Use Policy 107:105456

24. Wang Q, Chen T T, Li A D et al (2020) Evaluation of territorial utilization quality based on "production-living-ecological" perspectives—a case study of Panxi region. Mt Res 38(2):290-302
25. Yang H, Fang CP, Lin HL (2017) Research on the evaluation of efficiency in land spatial development and utilization at the county level in Fujian Province Chinapopulation. Resources and Environment 27(S1):109–113
26. Cai GY, Zhang JX, Du MY et al (2020) Identification of urban land use efficiency by indicator-SDG 11.3.1. Plos One 15(12):e0244318
27. Zhou ML, Lu LL, Guo HD et al (2021) Urban sprawl and changes in land-use efficiency in the Beijing-Tianjin-Hebei Region, China from 2000 to 2020: a spatiotemporal analysis using earth observation data. Remote Sens 13(15):2850
28. Kilkis S (2022) Urban emissions and land use efficiency scenarios towards effective climate mitigation in urban systems. Renew & Sustain Energy Rev 167:112733
29. Chakraborty S, Maity I, Dadashpoor H et al (2022) Building in or out? Examining urban expansion patterns and land use efficiency across the global sample of 466 cities with million plus inhabitants. Habitat Int 120:102503
30. Ma JF, Li WF, Wang Z et al (2022) Measuring multi-faceted land use efficiency of large-scale urban agglomerations under multi-scale drivers: Evidence from China. Land 11(1):91
31. Langpap C, Hascic I, Wu JJ (2008) Protecting watershed ecosystems through targeted local land use policies. Am J Agric Econ 90(3):684–700
32. Makowski D, Hendrix EMT, Van Ittersum MK et al (2000) A framework to study nearly optimal solutions of linear programming models developed for agricultural land use exploration. Ecol Model 131(1):65–77
33. Cao YG, Zhou W, Qiao LY et al (2012) Research on urban construction land changes and using benefit in eastern China. Prog Geogr 31(7):869–877
34. Zhang L, Zhang L, Xu Y et al (2020) Evaluating urban land use efficiency with interacting criteria: An empirical study of cities in Jiangsu China. Land Use Policy 90:104292
35. Aigner D, Lovell CAK, Schmidt P (2023) Formulation and estimation of stochastic frontier production function models? J Econ 234(SI):15–24
36. Den XZ, Gibson O (2020) Sustainable land use management for improving land eco-efficiency: A case study of Hebei, China. Ann Oper Res 290(1-2):265–277
37. Jin G, Deng XZ, Zhao XD (2018) Spatio-temporal patterns of urban land use efficiency in the Yangtze River Economic Zone during 2005-2014. Acta Geogr Sin 73(7):1242–1252
38. Ding L, Bai YP, Zhang LJ et al (2023) Spatio-temporal evolution of land use and eco-efficiency in urban agglomerations in the Yellow River Basin. Acta Ecol Sin 43(18):7610–7622
39. Attardi R, Cerreta M, Sannicandro V et al (2018) Non-compensatory composite indicators for the evaluation of urban planning policy: the land-use policy efficiency index (LUPEI). Eur J Oper Res 264(2):491–507
40. Cook WD, Seiford LM (2009) Data envelopment analysis (DEA)—thirty years on. Eur J Oper Res 192(1):1–17
41. Jiang WY, Cai YL, Tian JJ (2019) The application of minimum cumulative resistance model in the evaluation of urban ecological land use efficiency. Arab J Geosci 12(23):714
42. Liu JS, Lu LYY, Lu WM et al (2013) Data envelopment analysis 1978–2010: a citation-based literature survey. Omega-Int J Manag Sci 41(1):3–15
43. Cook WD, Tone K, Zhu J (2014) Data envelopment analysis: prior to choosing a model. Omega-Int J Manag Sci 44:1–4
44. Liu JS, Lu LYY, Lu WM (2016) Research fronts in data envelopment analysis. Omega-Int J Manag Sci 58:33–45
45. Cook WD (2001) Data envelopment analysis: a comprehensive text with models, applications, references and DEA-solver software. Interfaces 31(3):116–118
46. Lu XH, Chen DL, Kuang B et al (2020) Is high-tech zone a policy trap or a growth drive? Insights from the perspective of urban land use efficiency. Land Use Policy 95:104583
47. Bovet J, Reese M, Köck W (2018) Taming expansive land use dynamics—sustainable land use regulation and urban sprawl in a comparative perspective. Land Use Policy 77:837–845

48. Chen X, Yu L, Du ZR et al (2022) Toward sustainable land use in China: a perspective on China's national land surveys. Land Use Policy 123:106428
49. Wang J, Chen YQ, Shao XM et al (2012) Land-use changes and policy dimension driving forces in China: present, trend and future. Land Use Policy 29(4):737–749
50. Bounouh O, Essid H, Farah I R (2017) Prediction of land use/land cover change methods: a study. 2017 3rd International Conference on Advanced Technologies for Signal and Image Processing (Atsip), 264–270
51. Kou J, Wang JJ, Ding JL et al (2023) Spatial simulation and prediction of land use/land cover in the transnational Ili-Balkhash Basin. Remote Sens 15(12):3059
52. Gashaw T, Tulu T, Argaw M et al (2018) Estimating the impacts of land use/land cover changes on ecosystem service values: the case of the Andassa watershed in the Upper Blue Nile basin of Ethiopia. Ecosyst Serv 31:219–228
53. Zhou L, Dang XW, Sun QK et al (2020) Multi-scenario simulation of urban land change in Shanghai by random forest and CA-Markov model. Sustain Cities Soc 55:102045
54. Uddin MS, Mahalder B, Mahalder D (2023) Assessment of land use land cover changes and future predictions using CA-ANN simulation for Gazipur City Corporation. Bangladesh Sustain 15(16):12329
55. Hajjar SE, Abdallah F, Kassem H et al (2023) Long short-term memory and attention models for simulating urban densification. Sustain Cities Soc 98:104852
56. Zhang P, Liu L, Yang LW et al (2023) Exploring the response of ecosystem service value to land use changes under multiple scenarios coupling a mixed-cell cellular automata model and system dynamics model in Xi an. China Ecol Indic 147:110009
57. Verburg PH, Van Berkel DB, Van Doorn AM et al (2010) Trajectories of land use change in Europe: a model-based exploration of rural futures. Landsc Ecol 25(2):217–232
58. Lu LL, Qureshi S, Li QT et al (2022) Monitoring and projecting sustainable transitions in urban land use using remote sensing and scenario-based modelling in a coastal megacity. Ocean & Coast Manag 224:106201
59. Liang J F (2021) Research on use efficiency of construction land and the drive factors in Fujian Province. Dissertation, Fujian Normal University
60. Hao S (2018) Study on the present evaluation and trend prediction for eco-efficiency in mainland China. Dissertation, Liaoning Normal University

Chapter 3
Indicator Selection: Construction of Geographic Factor Analysis Model

Abstract This chapter focuses on indicator selection and constructs the GeoFA model to address the spatial defects of the traditional FA model in TSUE research. It first analyzes the core logic of the traditional FA model and its prominent spatial blind spots, namely neglecting spatial dependence and spatial heterogeneity. Then, it elaborates on the theoretical connotation of geospatial effects and the theoretical challenges posed by spatial non-stationarity to traditional models. On this basis, two technical paths for the GeoFA model construction are proposed: one based on the geographically weighted correlation coefficient matrix and the other on the geographically weighted spatial similarity matrix. Their mathematical expressions and weight matrix construction methods are also provided. Finally, multi-model comparative experiments are conducted to evaluate the two proposed paths. The GeoFA model based on the geographically weighted correlation coefficient matrix is selected as the optimal model. This model provides scientific and technical support for indicator dimensionality reduction in TSUE evaluation.

3.1 Model Evolution and Theoretical Basis: From Traditional Factor Analysis Model to Geographic Factor Analysis Model

3.1.1 Core Logic and Spatial Limitations of the Traditional FA Model

Factor analysis serves as a classic multivariate statistical dimension reduction method. Its core function lies in eliminating data redundancy among high-dimensional observed variables [1]. This method explores potential and unobservable common factors behind original variables. It condenses core data information and delivers concise and interpretable low-dimensional features for subsequent analysis. The traditional FA model decomposes all observed variables into common factors

Y. Ma, *Evaluation and Prediction Modeling of Territorial Space Use Efficiency*, Urban Sustainability, https://doi.org/10.1007/978-981-92-4396-9_3

and specific factors [2]. It distinguishes common correlations and individual differences among variables and simplifies the covariance structure of high-dimensional data. Data dimension reduction can be realized with the support of a small number of core factors. Reliable theoretical logic enables the wide application of this method in non-spatial research fields such as psychology and sociology.

Global assumptions lay the theoretical foundation of the traditional FA model and form the core limitation for its application to spatial data [1, 2]. Such assumptions run through the whole model operation. All samples are assumed to follow a unified generation rule. Overall correlation can effectively reflect variable characteristics, and global factor loadings maintain universal applicability. The model also sets independent relationships between common factors, specific factors, and various internal factors. The traditional FA model completely ignores the spatial attributes of research samples. It neglects the influence of geographical location on variable correlation rules and factor action paths. Such basic settings create obvious contradictions with the core theories of spatial econometrics.

From the perspective of spatial econometrics, the global framework of the traditional FA model leads to prominent spatial defects. This method fails to fully consider spatial heterogeneity and spatial dependence. Core inherent characteristics of geographic data cannot be accurately identified, which further causes systematic errors in final research conclusions.

The neglect of spatial dependence further weakens the spatial applicability of the traditional FA model. Spatial dependence represents a typical feature of geographic data [3]. Observed values of a single spatial unit are closely affected by adjacent units. This feature widely exists in various territorial space data and ecological environment data. It also acts as a key distinction between spatial statistics and traditional multivariate statistics. The traditional FA model takes sample independence as a basic premise. This setting conflicts with the spatial interaction characteristics of geographic elements. In the stage of parameter estimation, spatial correlation between units is wrongly classified as inherent correlation of variables themselves. Such bias leads to inaccurate calculation results of factor loadings. This model cannot effectively distinguish internal variable correlation and external correlation caused by geographic proximity. Extracted common factors are highly likely to contain invalid interactive information. In the research of air quality, the spatial interaction of pollutants among adjacent cities is simply attributed to unified pollution driving factors. The explanatory capacity of common factors will be overestimated, and the objectivity of relevant research conclusions will be weakened.

3.1.2 Theoretical Connotation of Geospatial Effects: Spatial Dependence and Spatial Heterogeneity

Spatial dependence derives from the First Law of Geography. This classic theory confirms universal correlations among all geographical objects, and strength of correlation increases with shorter spatial distance. No fully independent sample exists in geographical space [4]. Attribute features of each spatial unit are affected by neighboring units, and such influence gradually weakens with rising spatial distance. This process forms general spatial autocorrelation. Spatial autocorrelation falls into two basic types including positive spatial correlation and negative spatial correlation [5]. Positive spatial correlation reflects similar attributes among adjacent units, while negative spatial correlation embodies opposite attributes across neighboring spaces. Positive spatial correlation appears more frequently in research fields such as territorial space use, ecological environment management and economic development. Typical manifestations include similar industrial structures within urban agglomerations and consistent water pollution levels across regions in the same river basin.

Spatial heterogeneity summarizes the core implication of the Second Law of Geography and provides the theoretical basis of spatial non-stationarity [6]. Distribution patterns of geographical elements, interactive relationships between variables, and systematic operation rules change along with geographical locations. Spatial units differ inherently in natural conditions, human contexts, and socio-economic levels [7]. These inherent distinctions lead to obvious spatial differentiation in core driving factors and their action intensity behind geographical phenomena. Such spatial differentiation follows internal rules and directly reflects the complexity of geographical systems. Obvious differences can be observed in the action mechanisms and influence intensities between economic driving factors of eastern and central-western regions, as well as between ecological influencing factors of mountainous and plain areas.

Spatial dependence and spatial heterogeneity work jointly rather than independently in the research of TSUE [8]. Spatial dependence breaks the independence of efficiency performance and factor allocation across geographical units. Obvious factor spillover effects and coordinated development connections commonly exist among adjacent regions. Spatial heterogeneity denies the unified global standard of core driving factors and their contribution intensity for efficiency improvement. Regional differentiation displays objective inevitability in practical spatial research. These two characteristics constitute the fundamental attributes of geospatial data [9]. They also define clear theoretical boundaries for the application of traditional statistical models in TSUE research.

3.1.3 Theoretical Challenges and Adaptation Requirements of Spatial Non-stationarity Theory for Traditional Models

Spatial non-stationarity is the core external manifestation of spatial heterogeneity. It also reflects how spatial heterogeneity is applied in model research. Meanwhile, it acts as the essential theoretical cause for the limited applicability of the traditional FA model in geographical research. The core implication of spatial non-stationarity indicates that statistical characteristics of geographical variables, correlation between variables, and internal driving mechanisms change significantly with spatial locations. No unified and stable functional form exists across the entire research area. This theory fundamentally challenges the core assumptions of the traditional FA model and brings systematic theoretical challenges.

Spatial non-stationarity directly challenges the general unique structure assumption of the traditional FA model. The traditional FA model adopts fixed settings across the whole study area. Such methods regard all spatial samples as a single homogeneous group and ignore the complexity and differentiation of geographical systems. Regional differences widely exist in natural background conditions, social and economic development levels, and spatial governance policies across practical geographical contexts. Core driving factors of geographical phenomena and their influencing intensities present evident spatial differentiation. Global and unified model settings fail to capture local spatial characteristics. Such defects lead to biased factor extraction and weakened model explanatory power and research conclusions inconsistent with objective geographical rules.

The combination of spatial non-stationarity and spatial dependence further exacerbates structural deviations of the traditional FA model. Spatial dependence exists as a universal and objective feature in geographical systems. Sample attributes of adjacent spatial units show strong correlation. Spatial non-stationarity leads to regional differentiation in such correlations. The traditional FA model cannot account for spatial correlation among samples or adapt to the differentiated correlation structures of variables. The superposition of two spatial features magnifies inherent limitations of traditional models. Common defects include biased estimation of factor loading, distorted extraction of common factors, and inaccurate identification of core driving factors. Traditional methods cannot meet the demands of dimensionality reduction and indicator selection in geographical spatial analysis.

The traditional FA model requires targeted optimization and improvement to solve existing theoretical and practical limitations. The overall improvement objectives include two key aspects. The first objective lies in breaking the constraints of global unified assumptions. Research shall abandon the single global factor solution and achieve localized representation of factor structures to adapt to general spatial non-stationarity. The second objective focuses on integrating spatial location information, adjacent spatial relationships, and spatial heterogeneity into model construction. The reconstruction of theoretical foundations helps capture dual characteristics of spatial dependence and spatial differentiation. Revised models can align theoretical assumptions with objective geographical laws. The optimized methods can provide solid technical and theoretical support for factor analysis in multi-scenario geographical research.

3.1.4 Design Concepts of the GeoFA Model: From Global Uniformity to Local Spatial Adaptation

The core optimization logic of the GeoFA model overcomes inherent limitations of the globally stationary factor structure within the traditional FA model. This model discards the single paradigm of global factor decomposition and forms a theoretical framework with local spatial adaptation as the core. By fully incorporating geospatial effects, the model remedies the essential defect of the traditional FA model in neglecting spatial attribute constraints. Current research takes geographic spatial units as basic research carriers. It captures variations in environmental conditions and data characteristics across locations and adjacent areas. Differentiated local factor structures can therefore be derived. Dynamic spatial evolution of factor loadings and variable correlation relationships also becomes achievable. Such theoretical design well accommodates spatial non-stationarity and spatial dependence inherent in geographic data. Analytical fitting precision and geographical explanatory capacity can be greatly enhanced for multi-scale spatial factor analysis.

The GeoFA model follows the basic logical framework of the traditional FA model at the technical level. Optimization with geographic weighting applies to the correlation coefficient matrix as the core computational component. Integration of spatial attenuation rules and neighborhood association constraints supports the establishment of a revised factor analysis method. This new method better adapts to multi-dimensional geographic indicator data. Two technical routes aim to resolve the common deficiency of the traditional FA model in ignoring spatial attribute constraints. Observable differences exist only in the quantitative measurement and characterization of spatial effects.

The first modeling route constructs the GeoFA model framework with a geographically weighted correlation coefficient matrix. It is built on the local fitting principle of geographically weighted regression. Spatial distance weights assist the modification of global correlation coefficients and support quantitative measurement of spatial effects through geographic proximity. Factor structures change gradually along with geographic locations under this route. Regional geographic differentiation characteristics can be fully reflected. This effectively eliminates analytical biases caused by inadequate spatial treatment in traditional models. The second modeling route adopts a geographically weighted spatial similarity correlation coefficient matrix for the GeoFA model construction. Introduction of spatial similarity criteria upgrades overall matrix operation logic. Spatial effect quantification depends on association similarity between geographic units. Accurate recognition of interaction relationships among neighboring units helps reduce redundant interference from spatial data. This methodological path effectively repairs systematic flaws induced by inadequate spatial constraints in traditional FA model.

3.2 Dual-path Modeling for Model Transition: Geographically Weighted and Spatial Similarity Matrices

3.2.1 Benchmark Model: Mathematical Formulation and Applicable Conditions of the Traditional FA Model

British psychologist Charles Spearman proposed the traditional FA model in 1904. This model has gained extensive application in multiple disciplinary fields. Typical research areas include psychology, management, economics, medicine, and geography. The traditional FA model serves as a multi-index statistical analysis method. It condenses numerous indicators with complex correlations into several independent and comprehensive common factors. This method can also be defined as a statistical modeling technology for extracting common factors from indicator groups. Common factors reflect the latent shared characteristics of original indicators. The basic principle of the traditional FA model lies in grouping indicators according to correlation degrees. Indicators within the same group present strong correlation. Indicators from different groups show weak correlation or no correlation. Each group corresponds to an independent basic structure, namely the common factor. Such modeling realizes the dimensionality reduction of multiple original indicators. A small number of uncorrelated and unobserved factors replace massive raw data. These derived factors retain the core information of original datasets and support further comprehensive analysis of object characteristics. The traditional FA model assumes the existence of latent factors behind all observable indicators. Latent factors control the formation and presentation of various indicators. Observable indicator variations are driven by these latent factor structures. The operational procedure of the FA model focuses on the identification and confirmation of these latent factors.

The traditional FA model includes two major categories. These categories include the exploratory FA (R-type) model and the confirmatory FA (Q-type) model. The exploratory FA model requires no prior theoretical assumptions. Researchers do not predefine the number of latent common factors before analysis. This model helps identify latent common factors and examine correlation patterns between indicators and measurement items. The confirmatory FA model sets clear assumptions about the quantity of latent common factors. Relevant structural models derive from mature theories or prior knowledge. This model verifies the consistency between theoretical structural models and actual research data. Differences in application prerequisites distinguish their research scenarios. The exploratory FA model applies to basic structure exploration, data simplification, and measurement scale development. The confirmatory FA model focuses on factor structure verification, hierarchical relationship analysis, and reliability and validity evaluation of measurement scales. The theoretical framework for research on TSUE remains in the initial exploration stage. Relevant technical specifications still lack systematic improvement. Research progress faces obvious shortages of effective prior knowledge. The present study therefore adopts the exploratory FA model for subsequent empirical analysis.

The traditional FA model establishes correlations between various indicators and common factors through mathematical construction. Let $X = (X_1, X_2, \ldots, X_p)$ represent a set of observable indicator data. This dataset contains a total of p independent indicators. The mathematical expressions for indicator data and common factors are presented in Eq. (3.1).

$$\left\{\begin{array}{l} X_1 = a_{11}F_1 + a_{12}F_2 + \cdots + a_{1m}F_m + \varepsilon_1 \\ X_2 = a_{21}F_1 + a_{22}F_2 + \cdots + a_{2m}F_m + \varepsilon_2 \\ \quad\quad\quad\quad \cdots\cdots \\ X_p = a_{p1}F_1 + a_{p2}F_2 + \cdots + a_{pm}F_m + \varepsilon_p \end{array}\right\}, \mathrm{m} < \mathrm{p} \tag{3.1}$$

In matrix form, the above equation is expressed in Eq. (3.2).

$$\begin{bmatrix} X_1 \\ X_2 \\ \vdots \\ X_p \end{bmatrix} = \begin{bmatrix} a_{11} & a_{12} & \ldots & a_{1m} \\ a_{21} & a_{22} & \ldots & a_{2m} \\ \vdots & \vdots & \vdots & \vdots \\ a_{p1} & a_{p1} & \ldots & a_{pm} \end{bmatrix} \begin{bmatrix} F_1 \\ F_2 \\ \vdots \\ F_m \end{bmatrix} + \begin{bmatrix} \varepsilon_1 \\ \varepsilon_2 \\ \vdots \\ \varepsilon_p \end{bmatrix} \tag{3.2}$$

F_1, F_2, …, F_m stand for common factors. $A = (a_{ij})_{p \times m}$ refers to the factor loading matrix. The statistical connotation of factor loading a_{ij} reflects the correlation coefficient between the *i*-th indicator and the *j*-th common factor. This coefficient quantifies the influence degree of each indicator on corresponding common factors. $\varepsilon = (\varepsilon_1, \varepsilon_2, \ldots, \varepsilon_p)$ represents specific factors. Such factors capture residual information that cannot be explained or covered by the m common factors.

The traditional FA model calculations can be completed based on Eqs. (3.1, 3.2) to extract common factors. The correlation coefficient matrix between paired indicators supports all model solving processes. The correlation coefficient is calculated as Eq. (3.3).

$$r_{xy} = \frac{Cov(x, y)}{\sqrt{D(x)}\sqrt{D(y)}} = \frac{\sum_{i=1}^{n}(x_i - \overline{x})(y_i - \overline{y})}{\sqrt{\sum_{i=1}^{n}(x_i - \overline{x})^2}\sqrt{\sum_{i=1}^{n}(y_i - \overline{y})^2}} \tag{3.3}$$

In this formula, r_{xy} represents the correlation coefficient between indicator x and indicator y. $Cov(x, y)$ denotes the covariance between the two indicators. $\sqrt{D(x)}$ and $\sqrt{D(y)}$ stand for the standard deviations of the two indicators respectively. $\overline{x}$ is the mean value of indicator x. $\overline{y}$ is the mean value of indicator y. n refers to the total number of samples.

3.2.2 Path 1: GeoFA Model with Geographically Weighted Correlation Coefficient Matrix

3.2.2.1 Model Principle: Spatial Non-stationarity and Locally Weighted Correlation Estimation

Geographic space cannot exist or evolve in isolation [3]. Numerous studies have emphasized and verified the necessity of incorporating spatial effects in geographic research and analysis. Such consideration corresponds to the unique attributes of geographic space. Spatial heterogeneity represents the most fundamental and core effect in geographic space. This concept encompasses both spatial form and spatial process implications. The uneven spatial distribution of geographic elements reflects the heterogeneity of spatial form. Typical elements include ground elevation, slope gradient, soil type, temperature conditions, and ecosystem types. Differentiated generation mechanisms and evolution rules appear for population structure, epidemic transmission, socio-economic output, and crop yield across various locations. Such distinctions embody the heterogeneity of spatial process, which is also defined as spatial non-stationarity. A variety of extended geographic spatial effects derive from the basic spatial heterogeneity. Common types include the proximity effect of spatial dependence, the distance decay effect of spatial interaction, and the scale effect of spatial zoning. Within the spatial scope of urban agglomerations and economic belts, joint development prevails among different cities. Cities with strong economic strength generate obvious radiation and driving influences. The output of resources, technologies, and human capital promotes the transformation and development of surrounding regions. Regions with shorter geographic distances to core cities receive more significant radiation and driving benefits. This phenomenon reflects typical proximity and distance-decay characteristics in geographic systems. Meanwhile, the manifestation and intensity of spatial interactions vary across different spatial scales. Such differences highlight the essential role of scale effect. These spatial effects shape the interactions among geographical elements, ecological conditions, and socio-economic systems. In summary, spatial effects must be fully considered in factor analysis dimensionality reduction and indicator selection for geographic and socio-economic indicators. Comprehensive consideration helps ensure the rationality and accuracy of relevant geographic research.

This study revises the calculation formula of the original correlation coefficient matrix on the basis of spatial heterogeneity. This revision supports the construction of the GeoFA model. Existing studies have widely introduced geographic spatial weights in geographic correlation analysis. Such methods fully incorporate various geographic spatial effects. From the perspectives of spatial form heterogeneity and spatial process heterogeneity, this paper analyzes the radiation and driving characteristics in socio-economic development across geographic spaces. The results indicate that regions with higher attribute values possess greater potential influence on surrounding geographic units. Targeted assignment of geographic spatial weights can effectively reflect such influence mechanisms. Weight values correspond to specific attribute magnitudes and quantify the relative importance of different geographic spaces. Schematic diagrams in the following section further illustrate the construction mechanism of the GeoFA model. As shown in → Fig. 3.1, specific rules for weight assignment follow clear logical constraints. First, Geographic Space A presents the maximum value of attribute i and ranks first within the entire region. This unit exerts the strongest influence on surrounding spaces and obtains the highest geographic spatial weight. Second, Geographic Space F ranks second in the regional value of attribute i. Its influence on neighboring units remains lower than that of Geographic Space A. This unit is assigned a relatively high spatial weight. Third, Geographic Space E, C, B, D, H, L, G, K, N and J record low values of attribute i. These units generate limited radiation effects and receive relatively small spatial weights. Fourth, Geographic Space I, O, P and M show the minimum values of attribute i across the study area. These units are given the lowest geographic spatial weights. The attribute i mentioned above can represent any measurement indicator covering economic, ecological, resource and other research dimensions. Correlation coefficients between weighted indicators change from the original R_{ij} to WR_{ij}. The geographically weighted correlation matrix is thereby calculated. Overall modeling procedures take specific regional indicators as the basic reference. Spatial weight values are determined by actual indicator magnitudes of each geographic unit. The final calculation of the geographically weighted correlation coefficient matrix realizes targeted improvement on the traditional FA model.

3.2.2.2 Spatial Weight Definition and Matrix Construction

The above mechanism analysis indicates that the construction of geographic spatial weights essentially quantifies the importance of each spatial unit within the regional system and its radiation influence on surrounding spaces. The value assignment logic and classification criteria directly determine the rationality and scientific rigor of spatial effect characterization. Existing research proves that no fixed standard framework exists for spatial weight determination. The selection of measurement methods, basic indicators, and classification standards should fully conform to research objects, research objectives, and data characteristics. Such principles ensure that the weight system can accurately reflect the inherent heterogeneity of geographic space. Guided by the research perspective on PLES use efficiency throughout this study, this section constructs a complete geographic spatial weight system suitable for the GeoFA model. The overall framework covers the functional dimensions of production, living, and ecological spaces.

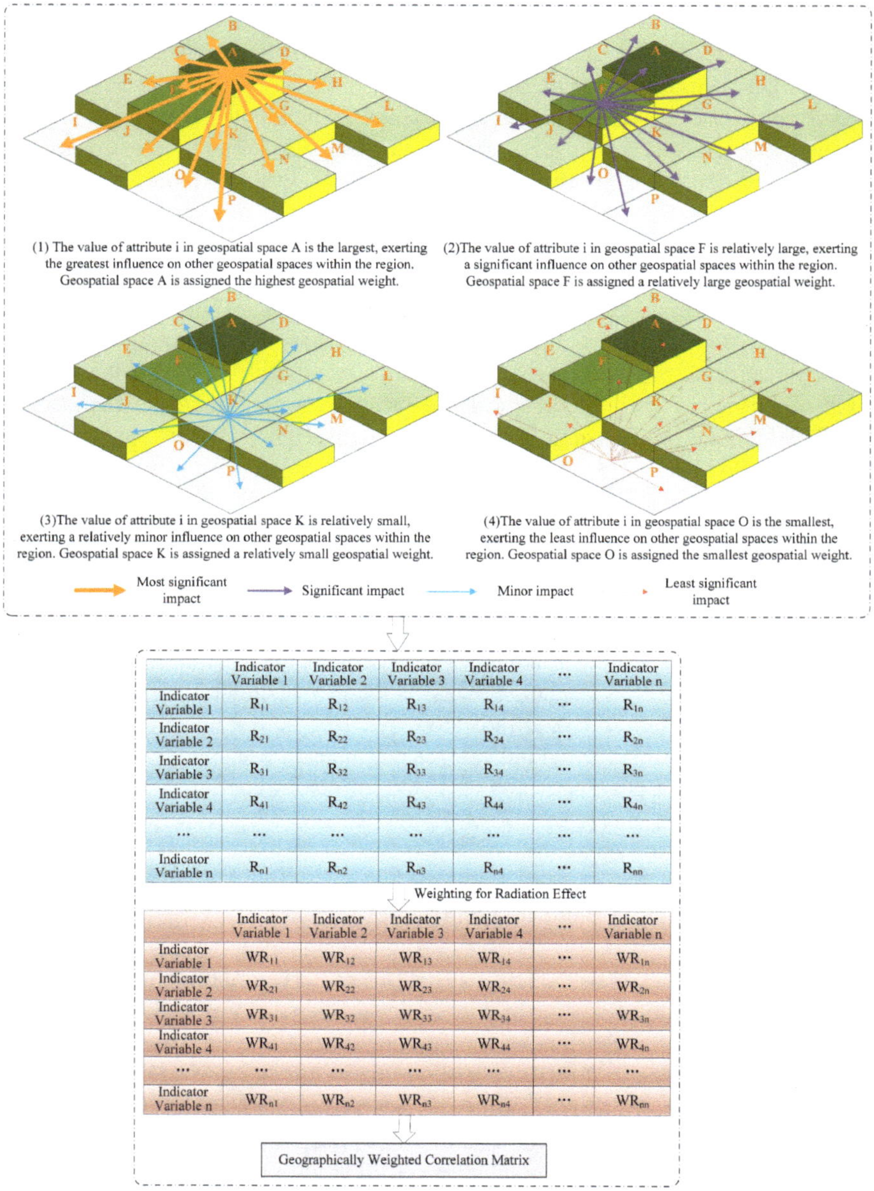

Fig. 3.1 Schematic diagram of traditional factor analysis model improvement based on geographically weighted correlation coefficient matrix

In terms of weight indicator selection, this study takes the scores of PLES functions as the core basis. Spatial function development level reflects the regional importance and spatial influence of each geographic unit. This design makes weight assignment consistent with the internal mechanism of TSUE. The specific calculation method for PLES function scores will be elaborated in → Chapter 6 of this study. This chapter only adopts the processed scores for weight assignment without detailed technical illustration. Geographic units show obvious differences in area size. Calculating spatial weights directly using raw function scores may introduce systematic errors arising from spatial scale heterogeneity. Large-sized units with low functional agglomeration may receive overestimated weights. Small-sized units with high functional activity may have their weights underestimated. Such defects will disturb subsequent spatial correlation analysis and factor extraction results. This study introduces the concept of per-unit average value to revise original function scores. The revision eliminates systematic deviations caused by area differences and reflects the actual functional intensity of each spatial unit more accurately. All revised values are further standardized through normalization and mapped to the range of [0,1]. Standardized results guarantee comparability across different geographic units and provide a stable and unified data foundation for subsequent weight classification and model operation.

For the classification of weight grades, this study adopts a multi-scheme comparative test to avoid one-sided conclusions from subjective setting. Three sets of equal-interval weight systems are established with three grades, five grades, and ten grades respectively. Comprehensive comparison is conducted from multiple dimensions. Key evaluation aspects include the quantitative distribution structure of territorial space, grade differentiation performance, and influences on the extraction results of common factors. Test results indicate that the three-grade classification contains excessive spatial units within each grade. Weak inter-grade differences fail to reflect the gradient characteristics of spatial importance. The ten-grade classification involves too many divisions. Some grades lack corresponding spatial units and present highly discrete spatial distribution. Meanwhile, different classification schemes exert no significant influence on common factor extraction. Considering grade differentiation, spatial distribution balance, numerical stability, and model applicability comprehensively, this study selects the five-grade equal-interval classification for geographic spatial weight construction. Scores in the range [0, 0.2) correspond to Weight 1. Scores in the range [0.2, 0.4) correspond to Weight 2. Scores in the range [0.4, 0.6) correspond to Weight 3. Scores in the range [0.6, 0.8) correspond to Weight 4. Scores in the range [0.8, 1.0] correspond to Weight 5.

In line with the principle of methodological consistency, the weight construction of living space and ecological space shares the same technical route and calculation logic as production space. Unified standards are maintained in indicator correction, normalization, and weight grading. This process forms a complete, standardized and logically consistent weight system for PLES. The systematic framework lays a solid methodological foundation for the construction of geographically weighted correlation coefficient matrices and the optimization of the GeoFA model.

3.2.2.3 Model Formulation and Estimation Method

Based on the above analysis and elaboration, the formula for the geographically weighted correlation coefficient matrix is constructed in Eq. (3.4).

$$r_{xy\_w} = \frac{\sum_{i=1}^{n}[w_i(x_i - \overline{x})][w_i(y_i - \overline{y})]}{\sqrt{\sum_{i=1}^{n}(w_i(x_i - \overline{x}))^2}\sqrt{\sum_{i=1}^{n}(w_i(y_i - \overline{y}))^2}} \tag{3.4}$$

In this formula, r_{xy_w} denotes the spatial correlation coefficient between indicator variable x and indicator variable y. w_i represents the geographic spatial weight of spatial unit i. x_i refers to the specific value of indicator x in geographic unit i. $\overline{x}$ stands for the mean value of indicator x across all geographic units. y_i refers to the specific value of indicator y in geographic unit i. $\overline{y}$ stands for the mean value of indicator y across all geographic units. n indicates the total number of geographic spatial units.

Since the traditional correlation coefficient satisfies the value range of $-1 \leq r_{xy} \leq 1$, the improved geographically weighted correlation coefficient shall also comply with $-1 \leq r_{xy\_w} \leq 1$. Such constraint ensures the smooth implementation of subsequent calculation procedures in factor analysis. This study verifies the above value range based on the Cauchy–Schwarz inequality, and the detailed proof is presented as follows.

Proof:

First, a quadratic function is constructed in the form of Eq. (3.5).

$$\begin{aligned}
\mathrm{F(x)} =& (w_1a_1x + w_1b_1)^2 + (w_2a_2x + w_2b_2)^2 + \ldots + (w_na_nx + w_nb_n)^2 \\
=& \left[(w_1a_1)^2x^2 + 2w_1a_1w_1b_1x + (w_1b_1)^2\right] + \left[(w_2a_2)^2x^2 + 2w_2a_2w_2b_2x + (w_2b_2)^2\right] \\
&+ \left[(w_na_n)^2x^2 + 2w_na_nw_nb_nx + (w_nb_n)^2\right] = \left[(w_1a_1)^2 + (w_2a_2)^2 + \ldots + (w_na_n)^2\right]x^2 \\
&+ 2[w_1a_1w_1b_1 + w_2a_2w_2b_2 + \ldots + w_na_nw_nb_n]x + \left[(w_1b_1)^2 + (w_2b_2)^2 + \ldots + (w_nb_n)^2\right]
\end{aligned} \tag{3.5}$$

Given $\mathrm{F(x)} \geq 0$, $\left[(w_1a_1)^2 + (w_2a_2)^2 + \cdots + (w_na_n)^2\right] \geq 0$

Thus,

$$\begin{aligned}
\Delta =& b^2 - 4ac \leq 0 = 4(w_1a_1w_1b_1 + w_2a_2w_2b_2 + \cdots + w_na_nw_nb_n)^2 \\
&- 4\left[(w_1a_1)^2 + (w_2a_2)^2 + \cdots + (w_na_n)^2\right]\left[(w_1b_1)^2 + (w_2b_2)^2 + \cdots + (w_nb_n)^2\right] \\
&\leq 0 = \left(w_1^2a_1b_1 + w_2^2a_2b_2 + \cdots + w_n^2a_nb_n\right)^2 \\
&\leq \left[(w_1a_1)^2 + (w_2a_2)^2 + \cdots + (w_na_n)^2\right]\left[(w_1b_1)^2 + (w_2b_2)^2 + \cdots + (w_nb_n)^2\right]
\end{aligned} \tag{3.6}$$

Simplifying the above expression yields the result in Eq. (3.7).

$$\left(\sum_{i=1}^{n} {w_i}^2 a_i b_i\right)^2 \leq \sum_{i=1}^{n} (w_i a_i)^2 \times \sum_{i=1}^{n} (w_i b_i)^2 \tag{3.7}$$

Dividing both sides of the inequality by the right-hand term yields Eq. (3.8).

$$\frac{\left(\sum_{i=1}^{n} {w_i}^2 a_i b_i\right)^2}{\sum_{i=1}^{n} (w_i a_i)^2 \times \sum_{i=1}^{n} (w_i b_i)^2} \leq 1 \tag{3.8}$$

Taking the square root yields Eqs. (3.9, 3.10).

$$\left(\frac{\left(\sum_{i=1}^{n} {w_i}^2 a_i b_i\right)^2}{\sum_{i=1}^{n} (w_i a_i)^2 \times \sum_{i=1}^{n} (w_i b_i)^2}\right)^2 \leq 1 \tag{3.9}$$

$$-1 \leq \frac{\left(\sum_{i=1}^{n} {w_i}^2 a_i b_i\right)^2}{\sum_{i=1}^{n} (w_i a_i)^2 \times \sum_{i=1}^{n} (w_i b_i)^2} \leq 1 \tag{3.10}$$

This reasoning verifies that the value range of Eq. (3.4) falls within [−1, 1]. The improved coefficient meets standard data requirements and supports subsequent calculations. Relevant follow-up procedures include eigenvalue solution, eigenvector calculation, factor loading matrix construction, factor rotation, and factor scoring.

3.2.3 Path 2: GeoFA Model with Geographically Weighted Spatial Similarity Matrix

3.2.3.1 Model Principle: Spatial Optimization from Variable Correlation to Sample Similarity

This study proposes a second technical approach for the construction of the GeoFA model, which improves the original correlation coefficient matrix based on the spatial neighborhood effect and distance decay effect. The analysis of neighborhood and distance relations requires at least two geographic spatial units. Geographic distance weights and economic distance weights both capture pairwise interactions across spatial units. By contrast, the traditional FA model calculates the correlation coefficient matrix by concentrating on the relationships between different indicator variables within a single geographic space. Accordingly, spatial linkage weights cannot be directly embedded in the original formula. Such a limitation restricts the application of the above idea in constructing the GeoFA model of this section.

This study addresses the above limitation by introducing the spatial similarity theory. This research idea is inspired by the calculation formula of Global Moran's I. Spatial similarity is a fundamental form of spatial relation. Its comprehensive and complex characteristics bring difficulties to quantitative description and numerical measurement, and existing relevant studies remain fragmented and unsystematic. No unified and widely recognized definition of spatial similarity has been formed by far, whereas this theory has been widely applied and studied in geographic spatial research [10]. In general, spatial similarity refers to the degree of similarity between two spatial objects or spatial scenarios in one or multiple dimensions. Comparable contents cover shape, scale, quantity, location, various attribute values and spatial functions. Several scholars further argue that spatial similarity focuses more on the correlation of attributes among spatial units, rather than the spatial distribution characteristics of individual or grouped spatial objects. Supported by the above theoretical analysis, this study improves the traditional correlation coefficient matrix under the guidance of spatial similarity theory, with full consideration of spatial neighborhood effect and distance decay effect [11]. The overall technical framework includes two core steps. First, this study calculates the spatial similarity of a single indicator between a given geographic unit A and all other geographic units. The calculation incorporates geographic distance weights, or a combination of geographic distance weights and economic distance weights. Spatial similarity of all indicators across all geographic units can be obtained through iterative computation. Second, the spatial similarity results derived from the first step are adopted as basic data to establish the corresponding correlation coefficient matrix. On this basis, the GeoFA model is constructed with the geographically weighted spatial similarity matrix.

This study further illustrates the construction mechanism of the GeoFA model in this section through the schematic diagram presented in → Fig. 3.2. Assume that geographic unit A in the diagram contains four distinct indicator attributes, and two scenarios are discussed in this research. In the first scenario where only geographic distance is considered, geographic units sharing common edges or corners with unit A, namely G, H, B, C, D, E, and F, are strongly influenced by unit A and show strong spatial similarity. Geographic units with no shared boundaries or corners but relatively short distances to unit A, including I, J, L, U, R, N, and M, receive weaker influences and exhibit weaker spatial similarity. Units K, T, Q, P, and O, which are far from unit A, show negligible similarity. In the second scenario, both geographic distance and economic distance reflected by indicator attribute values are taken into account. If the four attribute values of geographic unit S show high spatial similarity to those of unit A, strong interaction can be identified between the two units in terms of indicator variables. Meanwhile, only two small-area geographic units lie between them. From the perspective of geographic distance, unit S is also significantly affected by unit A. According to the above analysis and improved calculation formulas, each indicator variable of the original geographic unit corresponds to a unique spatial similarity value, as shown in → Fig. 3.2. The correlation coefficient matrix is further calculated based on spatial similarity data, so as to complete the construction of the GeoFA model.

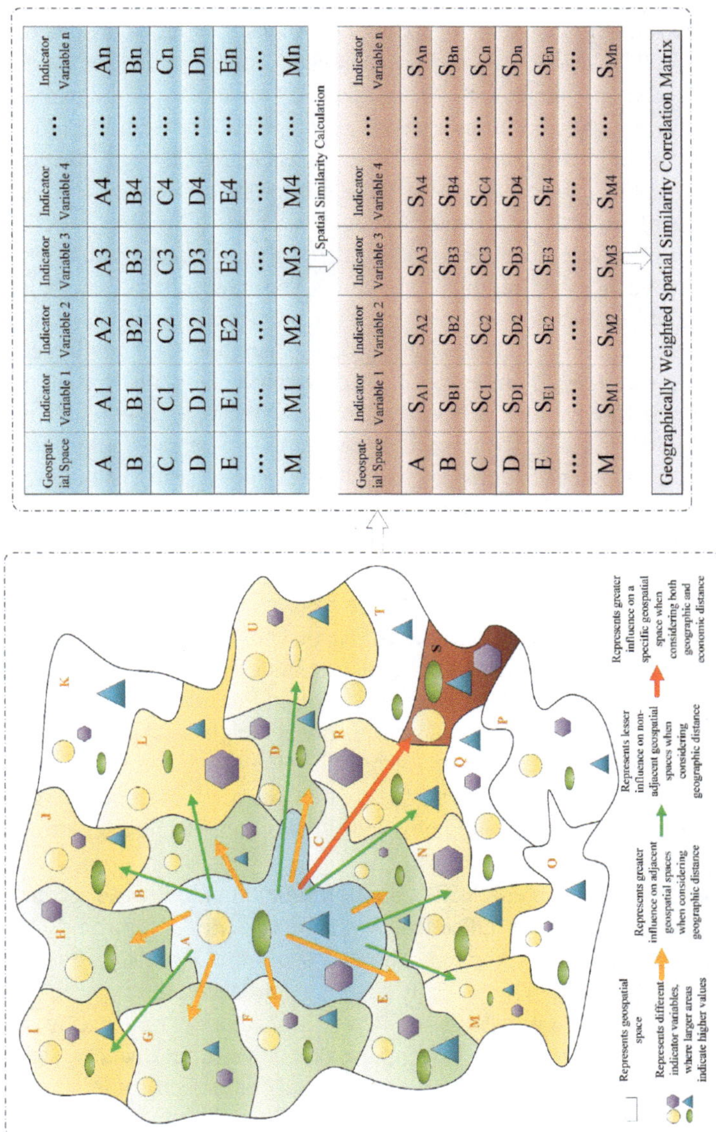

Fig. 3.2 Schematic diagram of traditional factor analysis model improvement based on geographically weighted spatial similarity correlation coefficient matrix

3.2.3.2 Spatial Similarity Measurement and Weight Matrix Construction

The weight adopted in this section is determined by the distance between pairwise geographic units. Based on the mechanism analysis in → Section 3.2.3.1, distance can be interpreted from two dimensions. The first dimension refers to distance in the geographic sense. The second dimension represents economic distance calculated through social and economic factors. Distance weight under the geographic dimension only quantifies locational correlation in geographical laws. Shorter distances correspond to stronger interactions, while longer distances weaken such correlations. Traditional geographic distance weights still present certain limitations. First, social, economic, and ecological development of geographic units have distinct economic attributes and urban development characteristics. Radiation effect and interconnection exert vital influences on regional operation. The fixed distance threshold of traditional geographic distance weights results in non-adjacency from the socio-economic perspective. For example, Beijing and Shanghai are both large cities with similar production and living patterns. The two cities show obvious adjacency features from the socio-economic perspective. However, the long geographic distance between the two cities excludes them from adjacent units in pure geographic measurement. Second, the symmetric setting of conventional geographic distance weights is inherently unreasonable. Geographic units with rapid economic growth and high output differ greatly from less developed units in radiation capacity. Such differentiation leads to obvious asymmetry in actual spatial interaction. Third, traditional geographic distance weight matrices lack time-varying characteristics. Such setting fails to conform to the evolving characteristics of geographic spaces. Under the conditions of available data, clear model structure, and definite influencing mechanism, the coupling of geographic distance weight and economic distance weight can form a combined geographic-economic distance weight matrix. To guarantee research comprehensiveness, this study constructs the GeoFA model separately based on geographic distance weight and integrated geographic-economic distance weight. Comparative analysis further helps select the optimal technical pathway for subsequent application.

First, this section clarifies the selection and formulation of traditional geographic distance weights. Such weights fall into two major categories according to adjacency rules and distance measurement. The two types differ greatly in judgment logic and operational complexity. Weight matrices constructed by adjacency rules hold clear principles and simple calculation steps. These matrices require no complex parameter configuration. By contrast, weight matrices based on distance measurement involve complicated theoretical mechanisms and cumbersome implementation procedures. Key parameters include radius thresholds and spatial distance indicators. These parameters depend on specific research contexts and require sufficient academic experience for reasonable calibration, which may introduce high subjectivity. The research framework of TSUE proposed in this study represents an innovative exploration. Relevant theoretical foundations and practical support remain insufficient. Arbitrary parameter setting may cause notable empirical biases. This study therefore adopts two concise and highly applicable weight specifications for GeoFA modeling. The selected tools include the binary Queen contiguity matrix and the distance power decay matrix. The geographic distance weight matrix W_g is established through ArcGIS Pro 3.0.

Secondly, this section specifies the formulation of geographic-economic distance weights. Existing studies widely adopt representative economic indicators to construct economic distance matrices as a mature empirical method. Economic indicators serve as valid measures to reflect the operational benefits of production space. These indicators also exert direct influences on resident income and consumption levels within living space. Regional economic development further exerts indirect effects on financial investment in environmental governance. Economic indicators exhibit direct and indirect correlations with production, living, and ecological spaces. This study therefore establishes the economic distance weight matrix with per capita GDP as the core indicator. The specific calculation formula is presented as Eq. (3.11).

$$M_{ij} = \frac{1}{\left|(PGDP_j - PGDP_i)/(PGDP_j + PGDP_i)\right|} \tag{3.11}$$

where M_{ij} denotes the economic distance between geographic unit i and geographic unit j. $PGDP_i$ and $PGDP_j$ represent the per capita GDP of geographic unit i and geographic unit j respectively. The economic distance weight matrix W_e can be further constructed accordingly.

Finally, this study adopts a multiplicative fusion method to integrate the geographic distance weight matrix and the economic distance weight matrix. The relevant calculation formula is expressed as Eq. (3.12).

$$W = W_g \times W_e \tag{3.12}$$

where W refers to the integrated geographic-economic distance weight, W_g represents the geographic distance weight, and W_e denotes the economic distance weight.

In the process of weight determination, this study designs four weight schemes according to actual research demands and data features to ensure research comprehensiveness. The detailed schemes are listed as follows. Scheme one applies only Queen-type geographic distance weight as the single weight. Scheme two adopts only distance power decay geographic distance weight as the single weight. Scheme three integrates Queen-type geographic distance weight with economic distance weight as the combined weight. Scheme four combines distance power decay geographic distance weight with economic distance weight as the combined weight.

3.2.3.3 Model Formulation and Solution Algorithm

The calculation formula of the correlation coefficient matrix based on geographically weighted spatial similarity is presented as Eq. (3.13).

$$s_{xy} = \frac{\sum_{a=1}^{n} h_{ax} \cdot h_{ay}}{\sqrt{\sum_{a=1}^{n} h_{ax}^{2}} \sqrt{\sum_{a=1}^{n} h_{ay}^{2}}} \tag{3.13}$$

where s_{xy} represents the spatial correlation coefficient matrix. h_{ax} denotes the spatial similarity of indicator x in geographic space a. h_{ay} refers to the spatial similarity of indicator y in geographic space a. The calculation formula for h_{ax} spatial similarity is presented as Eq. (3.14).

$$h_{ax} = \frac{n \sum_{b=1}^{n} w_{ab}(x_b - \overline{x})(x_a - \overline{x})}{\left(\sum_{b=1}^{n} w_{ab}\right) \sum_{a=1}^{n} (x_a - \overline{x})^2} \tag{3.14}$$

where w_{ab} indicates the weight between geographic space a and geographic space b. x_b represents the specific value of indicator variable x in geographic space b. x_a stands for the specific value of indicator variable x in geographic space a.

Further transformation and deduction are conducted for the above formula.

First, dividing the numerator and denominator in Eq. (3.14) by n respectively yields Eq. (3.15).

$$h_{ax} = \frac{\sum_{b=1}^{n} w_{ab}(x_b - \overline{x})(x_a - \overline{x})}{\left(\sum_{b=1}^{n} w_{ab}\right) \frac{\sum_{a=1}^{n} (x_a - \overline{x})^2}{n}} \tag{3.15}$$

Subsequently, dividing the numerator and denominator by $\left(\sum_{b=1}^{n} w_{ab}\right)$ separately yields Eq. (3.16).

$$h_{ax} = \frac{\frac{\sum_{b=1}^{n} w_{ab}(x_b - \overline{x})(x_a - \overline{x})}{\left(\sum_{b=1}^{n} w_{ab}\right)}}{\frac{\sum_{a=1}^{n} (x_a - \overline{x})^2}{n}} \tag{3.16}$$

The denominator of Eq. (3.16) represents the variance of indicator variable x, and the numerator reflects the weighted covariance of the data. Such formula design shares consistent logical principles with Moran's I. The two methods share inherent logical consistency. The calculation formula for the spatial similarity of h_{ay} yields Eq. (3.17).

$$h_{ay} = \frac{n \sum_{b=1}^{n} w_{ab}(y_b - \overline{y})(y_a - \overline{y})}{\left(\sum_{b=1}^{n} w_{ab}\right) \sum_{a=1}^{n} (y_a - \overline{y})^2} \tag{3.17}$$

The derivation of this formula aligns with that of h_{ax}. Relevant derivation steps are omitted in this section.

This study establishes the geographically weighted spatial similarity correlation coefficient matrix on the basis of (Eq. 3.13). The matrix substitutes the standard correlation coefficient matrix in the traditional FA model and completes the development of the GeoFA model. The proof for the value range of s_{xy} follows the research method presented in → Section 3.2.2.3. Detailed elaboration of this part is omitted.

3.3 Model Comparison and Empirical Selection

3.3.1 Comparative Experiment Design

To verify the superiority of the GeoFA model and select the optimal scheme, this study designs a series of multi-model comparison experiments. The core purpose lies in comparing the traditional FA model with the GeoFA models constructed through different pathways. These comparisons analyze performance differences in the dimensionality reduction of multi-dimensional geospatial indicators and identify the optimal model form. Specific details of the experimental design are presented below.

Experimental Environment: The GeoFA model is independently coded in Python 3.10 within the PyCharm 2023.2.1 Community Edition integrated development environment. Core dependency libraries include a data manipulation library including pandas 2.0.3, a scientific computing library including numpy 1.25.2, a plotting library including matplotlib 3.7.1, and a factor analysis library including factor_analyzer 0.3.2.

Experimental Subjects: This research adopts the candidate indicator system for living space use efficiency (LSUE) measurement established in the previous chapter as research samples. The indicator set contains 18 input indicators and 20 output indicators. Such selection guarantees the pertinence and scientific rigor of experimental data.

Experimental Variables: Model type is the core experimental variable. Six models are set for comparative analysis, including one traditional FA model and five GeoFA models with different construction pathways. Differences between models stem from the construction of correlation matrices and the setting of spatial weights.

Experimental Procedures: As shown in → Fig. 3.3, the model implementation and experimental execution include four steps. (1) Normalization and standardization of indicator data, KMO test, and Bartlett's test of sphericity are independently implemented based on core dependency libraries. These steps verify the suitability of data for factor analysis. (2) Calculation and visualization of two types of correlation matrices are performed. All programs are compiled according to Eq. (3.4, 3.13). The core modules of two types of the GeoFA models are constructed. (3) The core factor analysis library is invoked. Parameters are adjusted according to the GeoFA model design and research requirements. The principal component analysis method is adopted to estimate the loading matrix. The varimax method is used to achieve factor rotation of the loading matrix. Factor extraction and calculation for six models are completed. (4) Visualization libraries are applied to generate heat maps, line charts, and histograms for correlation coefficient matrices, eigenvalues, and primary factor scores. These figures directly present the model operation results. They provide support for the subsequent comparative analysis.

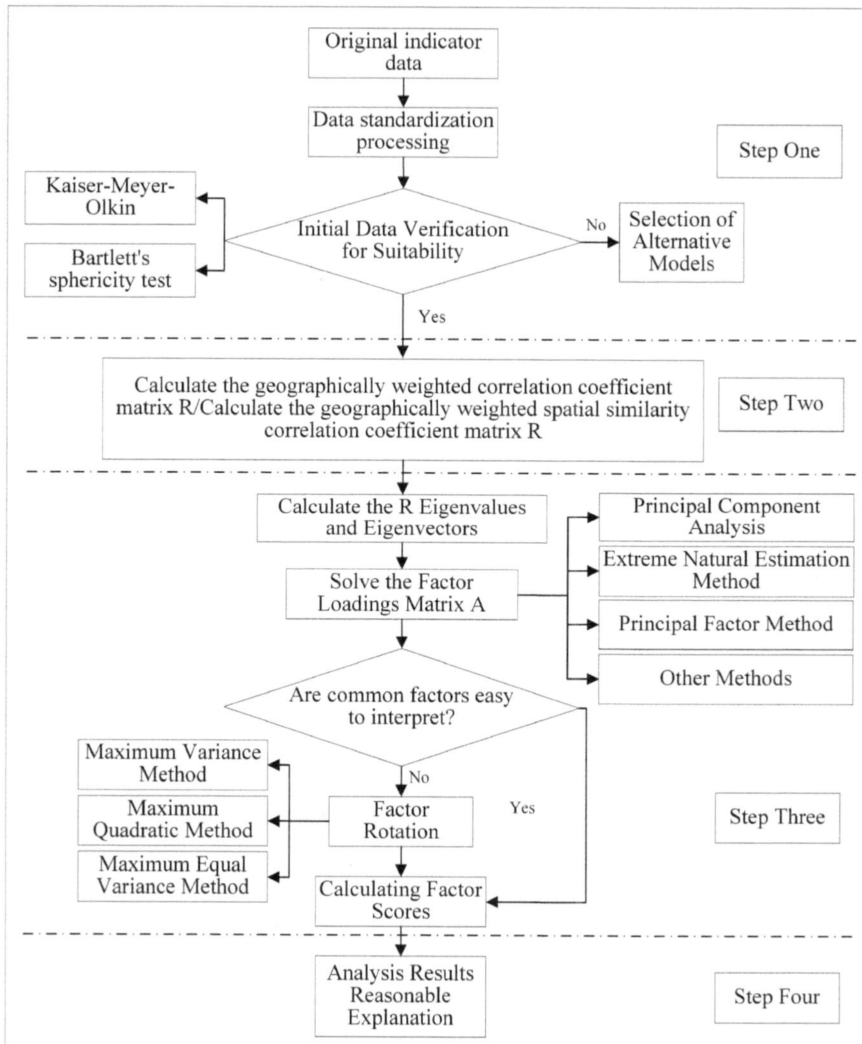

Fig. 3.3 Implementation process of the geographic factor analysis model

3.3.2 *Analysis of Empirical Results*

3.3.2.1 Model Verification System and Research Methods

As presented in → Table 3.1, this study formulates a multi-model comparison scheme for systematic verification on the applicability and superiority of the GeoFA model. The scheme takes LSUE as the typical research object. Comparisons with the traditional FA model help examine model performance under different weight structures and correlation coefficient matrix construction methods. This study finally determines the optimal model form suitable for the measurement of TSUE.

To ensure scientific and standardized verification procedures, this study adopts the previously established candidate indicator system for LSUE measurement as the fundamental basis for model testing. This candidate system integrates existing research findings, data accessibility and geospatial expression characteristics. The system contains 18 input indicators and 20 output indicators. Comparative factor analysis experiments of multiple models are carried out on the above basis.

This study conducts model verification from four dimensions, namely the KMO test, Bartlett's test of sphericity, common factor quantity, and cumulative contribution rate [11, 12]. The KMO test and Bartlett's test of sphericity complete the preliminary inspection of indicator data. The calculation formula of KMO is defined as Eq. (3.18).

$$KMO = \frac{\sum\sum_{i\neq j} r_{ij}^2}{\sum\sum_{i\neq j} r_{ij}^2 + \sum\sum_{i\neq j} \alpha_{ij.1,2\ldots,k}^2} \tag{3.18}$$

Table 3.1 Model types involved in verification and analysis

	Model	Model construction idea	Weight determination method
Model 1	Traditional FA Model	None	None
Model 2	GeoFA Model	GeoFA model constructed with geographically weighted correlation coefficient matrix	Geospatial weight considering the attribute value of geographic units
Model 3		GeoFA model constructed based on the geographically weighted spatial similarity correlation coefficient matrix	Queen-type geographic distance weight
Model 4			Distance power decay geographic distance weight
Model 5			Combination of Queen-type geographic distance weight and economic distance weight
Model 6			Combination of distance power decay geographic distance weight and economic distance weight

In this formula, r_{ij} refers to the correlation coefficient, and $\alpha^2_{ij.1,2...,k}$ represents the partial correlation coefficient. A *KMO* value higher than 0.9 indicates excellent suitability for factor analysis. Values between 0.8 and 0.9 reflect good suitability. Values ranging from 0.7 to 0.8 show acceptable suitability. Values within the range of 0.6 to 0.7 mean marginal suitability. Values between 0.5 and 0.6 reveal poor suitability. Data with a *KMO* value below 0.5 cannot meet the basic requirements for factor analysis.

Bartlett's test of sphericity evaluates the correlation among various indicators. The test sets a null hypothesis on the basis of the correlation coefficient matrix. A large test statistic and a p-value lower than 0.05 lead to the rejection of the null hypothesis. Data under such conditions satisfy the application conditions of factor analysis. In contrast, the data do not meet the requirements for factor analysis.

3.3.2.2 Empirical Analysis and Model Comparison

The number of common factors reflects the dimensionality reduction capacity of the model. The cumulative contribution rate reflects the explanatory power of factors for original data. Higher cumulative contribution rates correspond to better explanatory performance of the model.

Due to the large number of indicators and the wide temporal and spatial scope of this research, this section conducts detailed analysis with the candidate indicator system of LSUE. This arrangement clarifies the performance differences among various models. Separate multi-model comparisons are conducted on 18 input indicators and 20 output indicators. This study applies six models to carry out factor analysis on 18 input indicators. The relevant results are presented in → Fig. 3.4. The primary Y axis demonstrates the number of common factors. The secondary Y axis reflects the cumulative contribution rate and KMO test results. Bartlett's test of sphericity generates relatively uniform values for all samples. All models pass the test effectively. The relevant data are not displayed in the figure for concise presentation. The results of 1990 show consistent quantities of common factors across six models. Each model extracts five common factors. Obvious differences can be observed in cumulative contribution rates. Model 2 achieves the highest cumulative contribution rate. This outcome indicates stronger capacity to explain original data information. Meanwhile, all six models maintain favorable KMO test values. The data of 2000 present differentiated results in common factor extraction. Model 2 obtains six common factors. Model 1, Model 4 and Model 5 extract seven common factors. Model 3 and Model 6 extract eight common factors. Model 3 and Model 6 occupy the leading position in cumulative contribution rates. This advantage is closely related to the larger number of extracted common factors. Model 2 achieves a higher cumulative contribution rate than Models 1, 4 and 5. Model 2 maintains sufficient explanatory power for original data with fewer common factors. In 2010, only Model 2 extracts five common factors among all schemes. The remaining five models gain six common factors respectively. Model 2 ranks first in both cumulative contribution rate and KMO test value. Model 3, Model 4, Model 5 and Model

6 present better performance than Model 1 in the above two indicators. The operation results of 2020 share similar characteristics with the data of 2010. Model 2 extracts four common factors. Other models obtain five common factors. Model 2 still achieves the optimal values of cumulative contribution rate and KMO test. The corresponding indicators of the other five models remain at nearly identical levels.

This study applies six models to perform factor analysis on 20 output indicators. The relevant outcomes are presented in → Fig. 3.5, which contains the same evaluation indicators as → Fig. 3.4. The results of 1990 indicate that only Model 2 yields eight common factors. All other models generate nine common factors. Model 2 achieves the highest cumulative contribution rate under such conditions. This further shows that Model 2 can effectively interpret raw data with fewer common factors. In 2000, Model 2 extracts six common factors, while all other models extract eight common factors. Model 2 records relatively lower values in cumulative contribution rate and KMO test. All indicators still satisfy the normative standards and operational requirements for model application. The cumulative contribution rate of Model 2 is merely 0.0200 lower than that of other models with eight common factors. This slight gap further validates the high reliability of Model 2. The six models show consistent changing trends in 2010 and 2020. Model 2 extracts seven common factors in both periods, and the remaining models produce eight common factors. Model 2 still reaches the optimal level in cumulative contribution rate and KMO test results. Relevant quantitative indicators of the other five models remain roughly equal.

Through comprehensive evaluation, the GeoFA model constructed with the geographically weighted correlation coefficient matrix, namely Model 2, shows prominent advantages among all six schemes. This model realizes efficient dimensionality reduction and improves the explanatory capacity of original data. The research findings fully verify the superiority and practicality of the GeoFA model in dimensionality reduction for multi-dimensional geospatial indicators.

3.3.3 Comprehensive Conclusions

First, the GeoFA model has overall better performance than the traditional FA model. Empirical findings reveal that two types of the GeoFA models can effectively capture the spatial non-stationarity and spatial dependence of geospatial data. These models reduce factor extraction bias caused by non-spatial processing of traditional FA model. All the GeoFA models obtain evident improvements in core indicators such as cumulative contribution rate and KMO test results. These advantages are more prominent in regions with significant spatial heterogeneity. The research results fully confirm the rationality and necessity of optimizing traditional FA model by incorporating geospatial effects.

Second, the GeoFA model based on the geographically weighted correlation coefficient matrix, namely Model 2, achieves the optimal performance. This model presents prominent strengths in multi-temporal factor analysis of living space input and output indicators. The model realizes effective dimensionality reduction. The

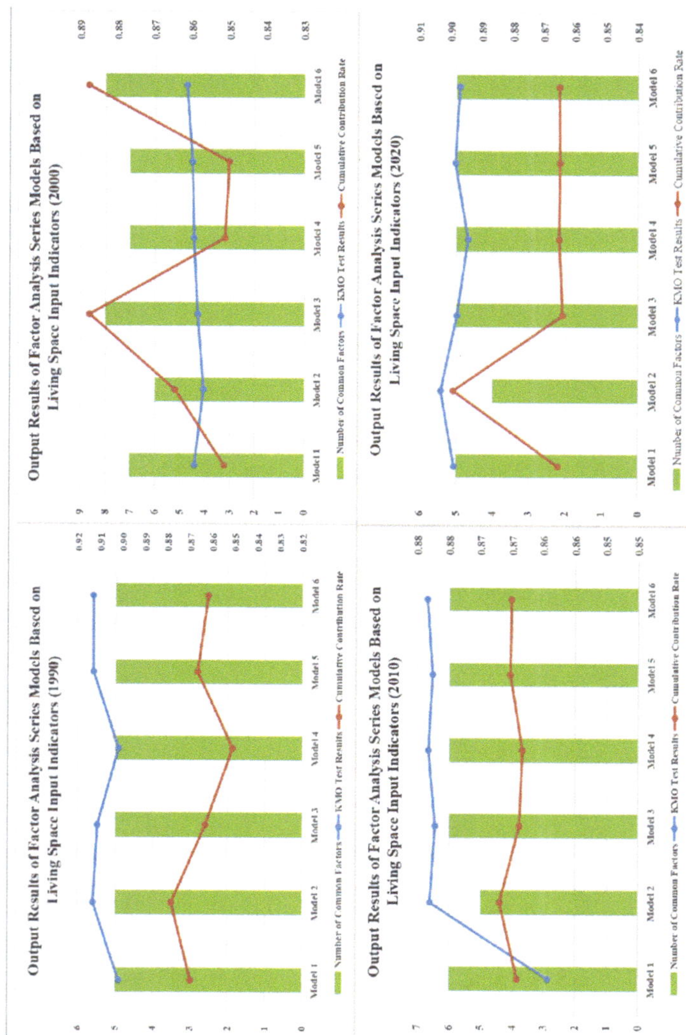

Fig. 3.4 Output results of series factor analysis models based on living space input indicators

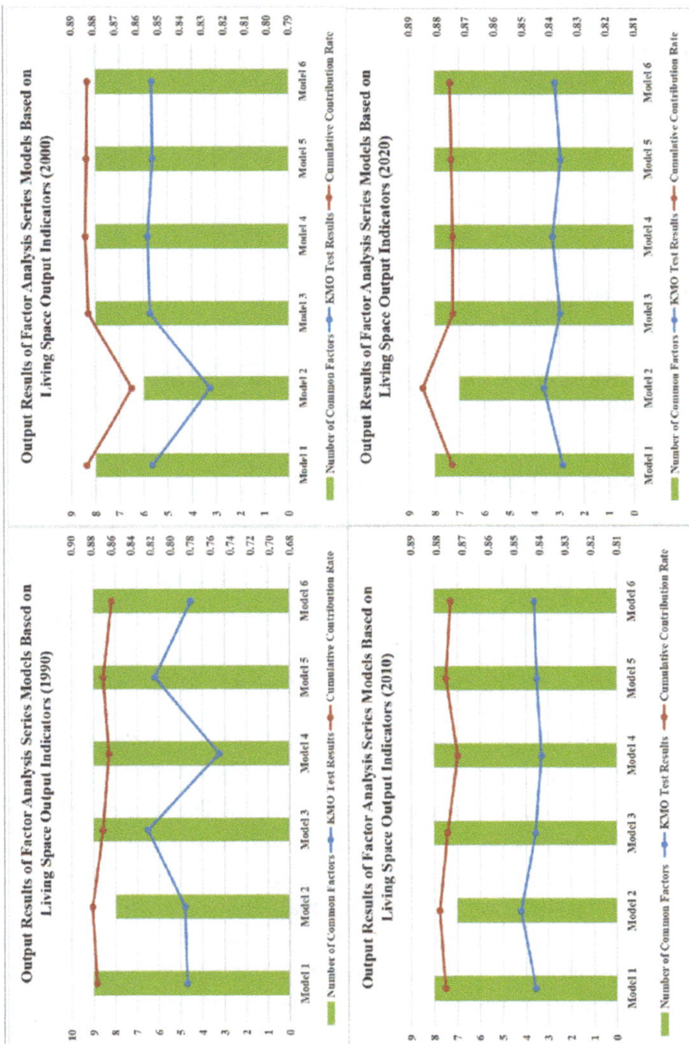

Fig. 3.5 Output results of series factor analysis models based on living space output indicators

number of common factors decreases by approximately 14% compared with the traditional FA model, and the overall structure of the indicator system gains effective simplification. The model improves explanatory capacity, lifting the total cumulative variance contribution by approximately 1.17%. and the core information of original indicators obtains sufficient retention in the process of dimensionality reduction. The model features high regional adaptability. Weights formulated according to spatial attribute values can accurately characterize spatial heterogeneity. The optimized factor structure can well match the actual geographic pattern of research regions. The model maintains stable computational performance. The modeling process excludes complex spatial similarity calculation and multi-weight coupling operation. The model is suitable for dimensionality reduction analysis for large-scale geospatial datasets with multiple indicators.

3.4 Chapter Summary

3.4.1 Methodological Contributions: A Full Evaluation Framework for the GeoFA Model

This chapter targets the dimensionality reduction demand of multi-dimensional geospatial indicators. The chapter completes the theoretical evolution and empirical verification from the traditional FA model to the GeoFA model. The core methodological contribution lies in the establishment of a complete GeoFA model evaluation framework. This framework follows a logical process of theoretical analysis, model construction, empirical comparison, and conclusion proposal. The research addresses methodological gaps of the traditional FA model in geospatial application scenarios.

The core value of this framework manifests in three key aspects. The framework defines the spatial limitations of the traditional FA model. It analyzes the theoretical challenges brought by spatial non-stationarity and spatial dependence. The analysis clarifies the specific direction for model optimization. The framework proposes two technical routes for the GeoFA model construction. It supports the transformation of factor analysis from global unified calculation to local adaptive simulation. The framework also establishes a multi-dimensional model evaluation system. The system delivers practical and measurable criteria for reasonable model selection.

This systematic framework solves the practical demand of indicator dimensionality reduction in current research. It also provides a referable methodological paradigm for similar studies with geospatial multi-indicator data. The research promotes the in-depth integration between traditional FA model and geospatial econometrics.

3.4.2 Applicable Scope and Notes on GeoFA Model

A clear definition of the applicable scope and application norms of the GeoFA model can improve the accuracy of model application. Clear definition also prevents improper or excessive model adoption.

In terms of applicable boundaries, the GeoFA model fits well with multi-indicator dimensionality reduction research at mesoscale and macroscale regions. Typical research contexts include provincial and municipal territorial space studies with obvious spatial non-stationarity and spatial dependence. The local adaptive features of the GeoFA model can be fully exerted under such research contexts. Traditional FA models remain sufficient for small homogeneous regions with weak spatial effects. Sufficient and high-quality spatial data constitute the basic prerequisite for stable GeoFA model operation. Researchers should exercise caution in model application under conditions of data deficiency or low data accuracy.

Four critical guidelines govern the practical implementation of this model. Reasonable configuration of spatial weight matrices relies on comparative analysis of multiple schemes. Comparative analysis helps identify the most suitable weight system for specific research contexts. Effective inspection of geographically weighted correlation coefficients guarantees smooth model calculation. All relevant coefficients should stay within the value range from -1 to 1. Comprehensive interpretation of model outcomes requires dynamic analysis across temporal and spatial dimensions. Multi-dimensional observation avoids one-sided research conclusions. Scientific explanation of analytical results needs solid support from geographic theories. Theoretical grounding ensures clear and reliable geographic implications for all research findings.

3.4.3 Future Outlook: From GeoFA Model to Complex Spatial Structure Modeling

The GeoFA model constructed in this chapter provides an effective approach for the dimensionality reduction of geospatial multi-indicator data. Combined with emerging research needs and technical trends, future research can carry out further expansion and improvement from four aspects.

First, future research can incorporate complex spatial features such as spatial scale effects and spatial interaction effects. The optimization of model structure can strengthen the adaptation capacity for complex geographic systems.

Second, integrated application of machine learning and deep learning methods can improve the accuracy of parameter estimation. Such integration can also expand the application scope of dimensionality reduction for high-dimensional spatial data.

Third, the research scope can extend to fields such as ecological environment evaluation and urban planning. Standardized calculation tools can be developed to support practical operation. Dynamic GeoFA models based on panel data can further enhance the practical value of relevant theories and methods.

Fourth, the introduction of temporal dimension can facilitate the construction of the spatial-temporal weighted factor analysis model. The new model can capture the temporal and spatial evolution rules of factor structures and offer solid support for dynamic research on regional development.

Competing Interests The authors have no conflicts of interest to declare that are relevant to the content of this chapter.

References

1. Moretti A, Shlomo N, Sakshaug JW (2020) Multivariate small area estimation of multidimensional latent economic well-being indicators. Int Stat Rev 88(1):1–28
2. Wang GJ, Zhao B, Wu BS et al (2023) Intelligent prediction of slope stability based on visual exploratory data analysis of 77 in situ cases. Int J Min Sci Technol 33(1):47–59
3. Liu Y, Wang KL, Xing XY et al (2023) On spatial effects in geographical analysis. Acta Geogr Sin 78(03):517–531
4. Walker RT (2022) Geography, VON Thunen, and Tobler's First Law: Tracing the evolution of a concept. Geogr Rev 112(4):591–607
5. Westlund H (2013) A brief history of time, space, and growth: Waldo Tobler's first law of geography revisited. Ann RegNal Sci 51(3):917–924
6. Foresman T, Luscombe R (2017) The second law of geography for a spatially enabled economy. Int J Digit Earth 10(10):979–995
7. Goodchild MF (2004) The validity and usefulness of laws in geographic information science and geography. Ann Assoc Am Geogr 94(4):709–714
8. Fotheringham AS, Yang WB, Kang W (2017) Multiscale Geographically Weighted Regression (MGWR). Ann Am Assoc Geogr 107(6):1247–1265
9. Wang JF, Li XH, Christakos G et al (2010) Geographical detectors-based health risk assessment and its application in the neural tube defects study in China. Int J Geogr Inf Sci 24(1):107–127
10. Yan HW (2010) Fundamental theories of spatial similarity relations in multi-scale map spaces. Chin Geogr Sci 20(1):18–22
11. Wang R, Yan HW, Lu XM (2022) Quantitative relations between topological similarity degree and map scale change of contour clusters in multi-scale map spaces. ISPRS Int J Geo-Inf 11(4):268
12. Ahmad US, Safdar S, Azam M (2024) An assessment of bilateral debt swap financing indispensable for economic growth and environment sustainability: a policy implication for heavily indebted countries. Environ Sci Pollut Res 31(4):5716–5734
13. Radovanovic M, Filipovic S, Panic AA (2021) Sustainable energy transition in central Asia: status and challenges. Energy Sustain Soc 11(1):4

Chapter 4
Efficiency Measurement: Construction of Geographic Data Envelopment Analysis Model

Abstract This chapter focuses on the measurement of TSUE and puts forward the GeoDEA model. First, it reviews the fundamentals of the traditional DEA model and highlights its flaw of ignoring geospatial effects. The discussion then elaborates on why spatial heterogeneity and spatial dependence should be incorporated into efficiency measurement. Integrating spatially constrained multivariate clustering with the generalized DEA model, the GeoDEA model reconstructs measurement sets and local efficiency frontiers. Using data from 340 Chinese cities for comparative analysis with the traditional DEA model, this study confirms that the GeoDEA model achieves higher accuracy, more appropriate reference frontiers, and outcomes in line with real urban development. The model serves as a dependable tool for TSUE evaluation.

4.1 Model Evolution and Theoretical Basis: From Traditional Data Envelopment Analysis to Geographic Data Envelopment Analysis Model

4.1.1 Traditional DEA Model: Basic Principles and Hypothetical Limitations

The DEA model represents a cross-disciplinary research paradigm covering management science, operational research, and economics. Charnes, Cooper, and other scholars first proposed this model in 1978 [1–3]. It has evolved over four decades. The academic community has proposed diverse understandings of the connotation and application of the DEA model.

Representative viewpoints from Banker, Wei, Zhou, and Liang have gained widespread academic recognition [4–7]. Core viewpoints summarized from existing studies can be classified into three key points. First, the DEA model serves as a

Y. Ma, *Evaluation and Prediction Modeling of Territorial Space Use Efficiency*, Urban Sustainability, https://doi.org/10.1007/978-981-92-4396-9_4

measurement tool for relative efficiency evaluation. Second, the modeling framework adopts non-parametric statistical techniques based on mathematical programming. Third, the analytical object of the DEA model refers to DMUs with homogeneous attributes. The model completes efficiency measurement for research units with multiple inputs and multiple outputs. All evaluated units need to possess unified economic attributes and similar operational characteristics [8].

The DEA model originates from production theory in microeconomics. Detailed investigation of production functions supports in-depth interpretation of the fundamental logic of the DEA model.

A production function describes the quantitative relationship between the quantity of various production factors and the maximum output. Such correlation holds for general production under specific periods and fixed technical conditions. A given production process involves two production factors of labor input (L) and capital input (K). Labor input serves as a variable factor, and capital input remains constant. The curves of total product of labor (TP_L), average product of labor (AP_L) and marginal product of labor (MP_L) in the production function are shown in Fig. 4.1. Production functions fall into three developmental stages based on the interaction between input factors and final output. Marginal product remains positive in the first and second stages. Positive marginal product reflects noticeable output growth along with the expansion of production input. Marginal product turns negative in the third stage. Additional input cannot bring output improvement at this stage. Rational production units always confine practical operation within a reasonable range and prevent production activities from entering the third stage.

The theoretical framework of production theory promotes the initial formation and subsequent systematic improvement of the DEA model.

Further taking the input-oriented BCC model as the basis, this section illustrates the basic logic of the DEA model with sample data. This study defines four DMUs, each containing one type of input (x) and one type of output (y), as listed in Table 4.1. The horizontal axis represents input (x) and the vertical axis represents output (y), as displayed in Fig. 4.2. Under the assumption of constant returns to scale, the

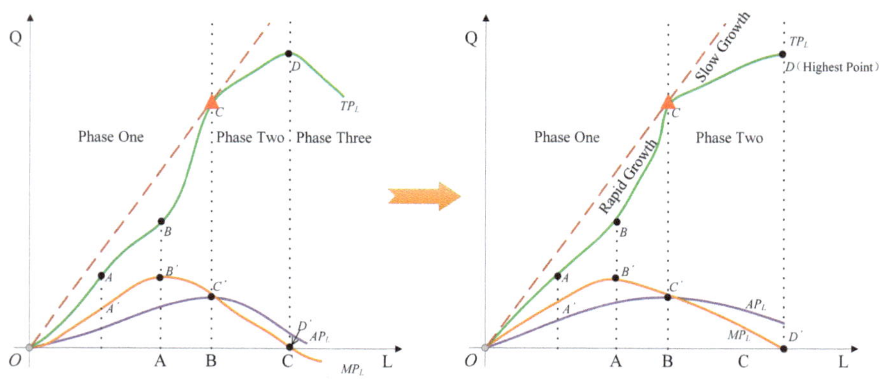

Fig. 4.1 Production function curve

production frontier corresponds to the ray connecting Point B and the origin. Point B represents the only effective DMU, while all other units are inefficient. Under the assumption of variable returns to scale, the production frontier forms a left convex curve composed of points M, A, B, and D. Point C refers to an ineffective DMU, and the input segment CC_{IV} reflects the level of input inefficiency. A clear comparison between the production function diagram of microeconomics (Fig. 4.1) and the DEA model diagram (Fig. 4.2) reveals high consistency between the production frontier and the curve or surface of total production output. This outcome further verifies that the theoretical development of the DEA model originates from production function theory. Relevant academic conclusions have also been confirmed by Tao and Yan [8]. Derived from such basic principles, a complete methodological system has gradually taken shape. The system includes basic the DEA models, diversified distance functions, super-efficiency SBM models, and has led to the proposal of the generalized DEA model concept. This study constructs the GeoDEA model based on the theoretical connotation of the generalized DEA model. The super-efficiency SBM model is adopted to calculate efficiency values throughout the measurement process.

This study fully incorporates the basic connotation and realistic background of territorial space use and incorporates the existence of undesirable outputs into the analytical framework. Most traditional models tend to classify most DMUs as efficient. Such results impede accurate ranking and comparative analysis of efficiency values. This research finally selects the distance function with the maximum distance to the frontier. The super-efficiency SBM model with undesirable outputs incorporation is adopted for efficiency calculation. The corresponding programming formula is presented as Eq. (4.1).

$$min\rho = \frac{1 + \frac{1}{m}\sum_{i=1}^{m} s_i^- / x_{ik}}{1 - \frac{1}{q_1+q_2}\left(\sum_{r=1}^{q_1} \frac{s_r^+}{y_{rk}} + \sum_{t=1}^{q_2} s_t^{b-} / b_{rk}\right)}$$

Table 4.1 Illustrative data

DMU	Input x	Output y
A	1.00	1.00
B	1.50	2.00
C	3.50	2.25
D	4.00	3.00

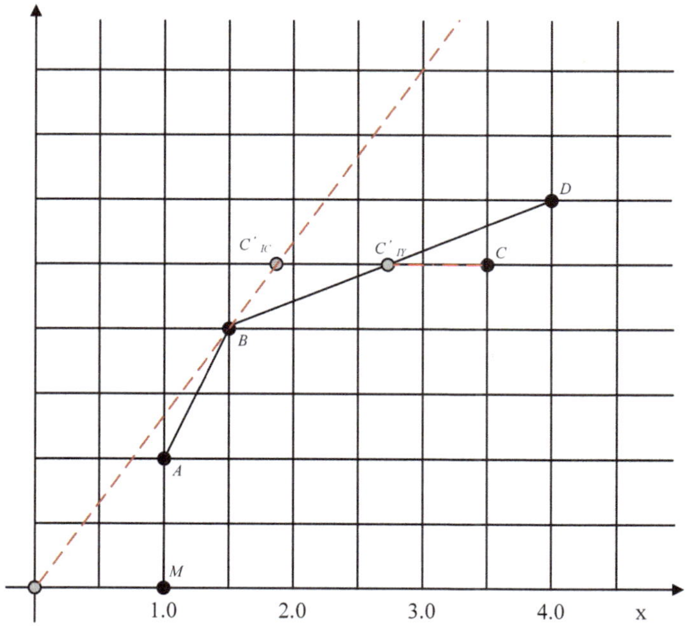

Fig. 4.2 Basic schematic diagram of input-oriented BCC model

$$
s.t.\left\{\begin{array}{l}
\sum_{j=1,j\neq k}^{n} x_{ij}\lambda_j - s_i^- \le x_{ik} \\
\sum_{j=1,j\neq k}^{n} y_{rj}\lambda_j + s_r^+ \ge y_{rk} \\
\sum_{j=1,j\neq k}^{n} b_{tj}\lambda_j - s_t^{b-} \le b_{rk} \\
1 - \frac{1}{q_1+q_2}\left(\sum_{r=1}^{q_1} \frac{s_r^+}{y_{rk}} + \sum_{t=1}^{q_2} s_t^{b-}/b_{rk}\right) > 0 \\
\lambda, s^-, s^+ \ge 0.
\end{array}\right\}
$$

$$
i = 1, 2, \ldots, m;\ r = 1, 2, \ldots, q;\ j = 1, 2, \ldots, n(j \neq k) \tag{4.1}
$$

The production possibility set consists of all DMUs except DMU k, as Eq. (4.2).

$$
\left\{(x, y) | x \ge \sum\nolimits_{j=1,j\neq k}^{n} x_{ij}\lambda_j,\ y \le \sum\nolimits_{j=1,j\neq k}^{n} y_{rj}\lambda_j\right\} \tag{4.2}
$$

Among these constraints, the expression $1 - \frac{1}{q_1+q_2}\left(\sum_{r=1}^{q^1} \frac{s_r^+}{y_{rk}} + \sum_{t=1}^{q^2} s_t^{b-}/b_{rk}\right) > 0$ can be eliminated during linear transformation. The symbol i denotes the category of input factors, with a total of m categories. The symbol r denotes the category of output factors, with a total of q categories. q_1 refers to desirable output variables, and q_2 refers to undesirable outputs. The symbol j represents DMUs, with a total of n units. The symbol k stands for the DMU under current evaluation. λ indicates the linear combination coefficient of DMUs. s_i^- represents the input slack variable. s_r^+ represents the slack variable of desirable outputs. s_t^{b-} represents the slack variable of undesirable outputs.

4.1.2 Geospatial Effects: Spatial Heterogeneity and Dependence in Efficiency Measurement

Classic DEA models have prominent limitations in the measurement of TSUE. The core constraint stems from its core underlying assumption of independent and identically distributed DMUs. This assumption completely ignores geospatial effects and deviates from the inherent spatial characteristics of territorial space use. This deviation is mainly reflected in two core dimensions, namely spatial heterogeneity and spatial dependence.

Spatial heterogeneity acts as a key factor that contradicts the fundamental assumptions of the traditional DEA models. According to the Second Law of Geography, the attribute characteristics and interactive relationships of geographic elements show systematic differences with the change of spatial locations [9]. Spatial units differ greatly in natural endowment conditions and socio-economic development levels. Such differences directly form inherent spatial gradients in the input-output efficiency of territorial space use [10]. Plain and mountainous areas, as well as eastern, central, and western regions, have distinct disparities in land input intensity, factor allocation efficiency, and output benefit structure. The traditional DEA model adopts a unified global production frontier for efficiency measurement. This method cannot reflect objectively existing spatial heterogeneity. It easily causes misjudgment of low-efficiency regions and overrating of inefficient units. Relevant measurement results will deviate from realistic conditions and lack practical explanatory power.

Spatial dependence further undermines the applicability of the traditional DEA models and serves essentially as the direct manifestation of the First Law of Geography in the field of territorial space utilization [11, 12]. TSUE exhibits obvious spatial spillover characteristics. The efficiency level of a single spatial unit depends not only on its own input-output allocation but also on direct and indirect influences from neighboring units in terms of industrial transfer, technology diffusion, and factor flow. Efficient development modes in core cities of urban agglomerations can generate radiation effects and drive the progress of surrounding regions. Meanwhile, extensive land use patterns may also bring negative impacts on adjacent units through spatial transmission. The traditional DEA models follow the premise of mutual independence among all DMUs. These models conduct independent evaluation merely based on data of individual units. Such practice ignores spatial transmission and interlinkage mechanisms of efficiency performance and fails to capture regional collaborative features under spatial spillover effects.

The objective existence of geospatial effects limits the capability of the traditional DEA models to achieve accurate measurement of TSUE. Scientific and effective efficiency evaluation requires breakthroughs in the inherent assumptions of the traditional DEA models. Spatial constraint mechanisms need to be introduced to regard geospatial characteristics as a basic prerequisite for efficiency measurement. The reconstruction of DMUs grouping logic and reference set selection mode promotes the transformation of efficiency evaluation from global unified assessment to local spatial adaptation. The above theoretical logic establishes the core theoretical foundation for the construction of the GeoDEA model.

4.1.3 Core Idea of GeoDEA Model: An Efficiency Evaluation Framework under Spatial Constraints

The core idea of the GeoDEA model is to overcome the inherent limitations of the traditional DEA models, which assume independent DMUs and a globally unified production frontier. This model integrates geospatial effects into the whole process of efficiency measurement and establishes an integrated framework of spatial constraint, cluster grouping, and efficiency evaluation. It thereby improves the geographical adaptability and accuracy of efficiency estimation. The core logic does not deny the non-parametric measurement advantages of the traditional DEA models. By introducing spatial constraints, the GeoDEA model corrects the applicability defects of the traditional DEA models in geospatial scenarios. The final efficiency results maintain rigorous mathematical logic and obtain clear geographical interpretation.

The core connotation of the GeoDEA model can be summarized into two key aspects. First, the model imposes spatial constraints in advance. This approach breaks the spatial homogeneity assumption embedded in the traditional DEA models and treats geospatial characteristics as a fundamental prerequisite for efficiency measurement. The traditional DEA models evaluate all DMUs based on a unified global production frontier. In contrast, the GeoDEA model fully incorporates spatial heterogeneity and spatial dependence. It adopts spatial clustering and other techniques to group DMUs with similar geographical features and tight spatial linkages. This grouping principle ensures high homogeneity within each group and reasonable spatial differentiation across different groups. Such a grouping strategy conforms to the inherent laws of geographical space. It provides a reliable reference foundation for subsequent efficiency evaluation and avoids efficiency misjudgments caused by blind comparisons of DMUs with inconsistent geographical conditions.

The second aspect lies in the localized reconstruction of production frontiers. Relevant research constructs local efficiency frontiers according to grouping outcomes and replaces the global frontiers applied in the traditional DEA models. Global frontiers formed by the traditional DEA models ignore spatial differences and cannot reflect differentiated efficiency features among various regions. The GeoDEA model establishes independent local frontiers for grouped DMUs with consistent geographical attributes. This design makes efficiency measurement results better match actual regional development conditions. DMUs in plain areas and mountainous areas can adopt independent local frontiers, respectively. This differentiated evaluation mode can accurately quantify the intensive utilization level of plain regions. It can also objectively assess efficiency performance of mountainous areas under ecological priority strategies. The overall evaluation system effectively improves geographical relevance of TSUE measurement.

In addition, the core thinking of the GeoDEA model is also reflected in the integration of spatial correlation. This model incorporates spatial dependence into the operating mechanism of efficiency measurement. The evaluation system focuses on the input and output allocation efficiency of individual DMUs. It also takes full account of spatial spillover effects between adjacent units to achieve comprehensive evaluation of independent efficiency and spatial linkage effects. This design retains the inherent strengths of the non-parametric method in the traditional DEA models. Such strengths include no requirement for preset production functions and strong adaptability to multi-input and multi-output scenarios. The introduction of spatial constraints addresses the poor geographical adaptability of traditional methods. The improved framework supports the accurate measurement of TSUE and lays a core theoretical foundation for subsequent model construction.

4.2 Geographic Data Envelopment Analysis: Spatial Constraints and Measurement Reconstruction

4.2.1 Model Construction Mechanism: Reconstruction of Measurement Sets and Introduction of Generalized DEA Model

This study adopts the generalized DEA model as the theoretical foundation and achieves the systematic integration of spatial constraints and efficiency measurement through measurement set reconstruction. The generalized DEA model is proposed by Ma et al. [13, 14]. Its core innovation lies in breaking through the limitation of unified measurement and reference sets in the traditional DEA models. In the traditional DEA models, the effectiveness of a single DMU is determined by comparison with the optimal units among all research samples. The generalized DEA model allows diverse mathematical matching relationships between the measurement set and the reference set. Common relationships include inclusion, intersection, and separation. Such settings enable seven flexible reference modes, such as self-comparison, cross comparison, and neighborhood group comparison. This feature provides a critical technical tool for the implementation of spatial constraints of the GeoDEA model. Spatial clustering technology partitions all DMUs into multiple homogeneous groups. Each group maintains internal homogeneity and reflects inter-group systematic differences with other groups. The differentiated correlation between measurement sets and reference sets can be established on the basis of grouping results. The whole process completes the logical realization of spatial constraint preposition and localized frontier reconstruction.

For special DMUs of territorial space, measurement set reconstruction requires comprehensive consideration of two categories of constraints. Geospatial constraints reflect the heterogeneous spatial distribution of basic geographic environments. Typical elements cover terrain, climate, and natural resources. Fragmented spatial configurations also widely exist in various socio-economic activities. Urban agglomerations and economic belts are typical manifestations of such distribution features. These realistic conditions define geographic proximity and environmental similarity as essential criteria for unit grouping. Socio-economic constraints rely on reasonable indicator selection to reflect actual development status. Common indicators include fixed asset investment, economic output, and population scale. Scientific indicator selection guarantees consistent efficiency formation conditions for units within the same group. The joint application of geospatial and socio-economic constraints conforms to basic geographic laws. This collaborative framework also meets the homogeneity standards of DMUs in standard DEA models. As Fig. 4.3 shows, the above logical content is presented in a visualized form.

The construction mechanism of the GeoDEA model can be summarized as follows. The model takes the generalized DEA model as the theoretical tool and adopts multi-dimensional clustering under spatial constraints as the technical method. It realizes the scientific reconstruction of measurement sets and achieves the goal of local efficiency measurement with comparable evaluation within homogeneous groups and consistent spatial regions.

4.2.2 Measurement Reconstruction: Division of DMU Sets Via Spatially Constrained Multi-Dimensional Clustering

Spatially constrained multivariate clustering represents a modeling technique for geographic analysis with spatial statistical methods. This technique performs clustering of geospatial elements under specific spatial constraints according to the attribute values of geographic elements and preset clustering quantities. This model requires no pre-classified elements for training or clustering guidance. Unsupervised machine learning algorithms dominate the determination of final clustering results. The model thus belongs to unsupervised machine learning methods. Under the restriction of preset clustering quantities, the spatially constrained multivariate clustering model seeks optimal solutions based on designated analysis fields. Relevant solutions ensure high similarity among elements inside each cluster and maximize inter-cluster differences. This model adopts connectivity graphs and minimum spanning trees as well as the SKATER method to complete the clustering of geospatial elements. The method acts as an effective exploratory tool. It supports in-depth interpretation of fundamental data structures and adapts well to clustering research that integrates

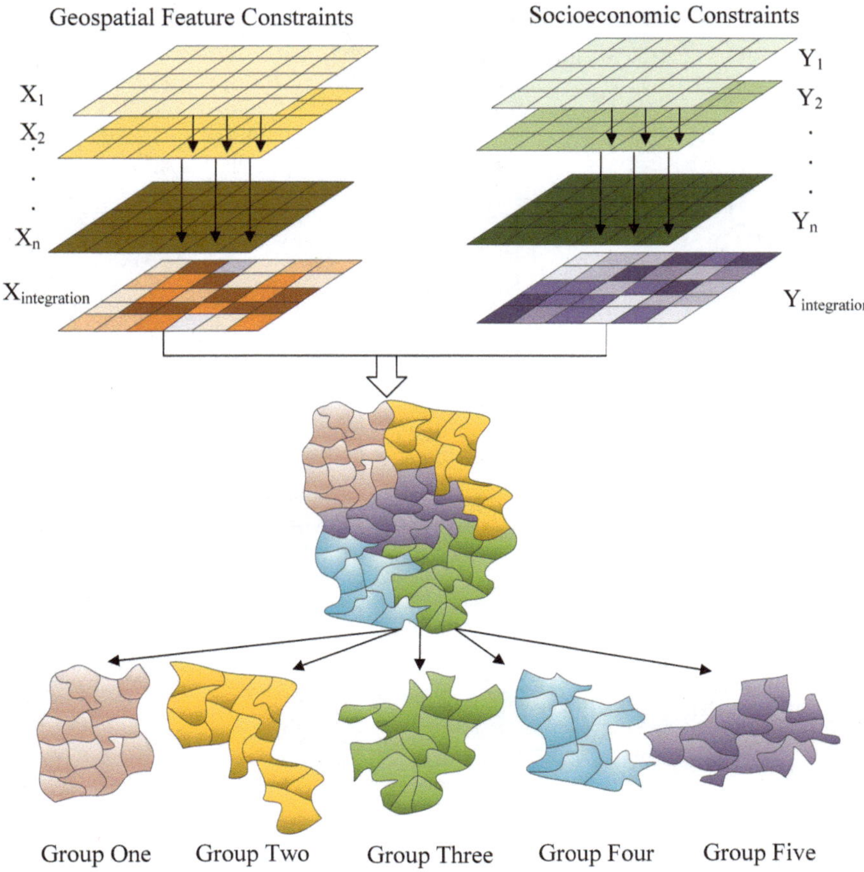

Fig. 4.3 Schematic diagram of measurement set reconstruction

geographic locations and indicator attribute data. This study applies the above model to territorial space clustering analysis. The model operation focuses on three core contents including spatial constraints, clustering quantity, and analysis fields. Analysis fields mainly refer to indicator attribute data without geographic characteristics. Figure 4.4 illustrates the relevant logical framework.

In terms of spatial constraints, relevant settings guarantee the spatial adjacency of all elements within a single cluster. Table 4.2 presents four major categories of mainstream spatial constraint methods. Spatial constraints of shared edges and shared edge corners apply only to cluster analysis of contiguous areal features. These two methods show limited applicability for areal features with isolated patches or discontinuous spatial distribution. The trimmed Delaunay triangulation constructs non-overlapping triangular networks based on feature centroids. Each feature acts as an independent triangular node. Nodes with shared edges are defined as neighboring units and can be grouped into the same cluster. Neighborhood boundaries

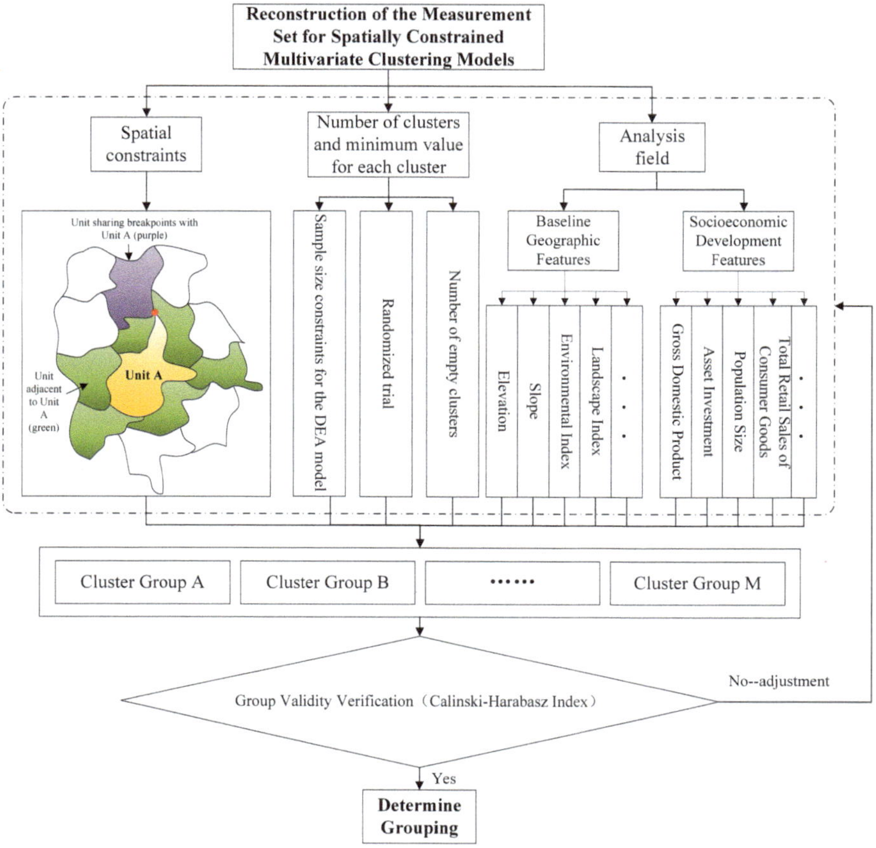

Fig. 4.4 Measurement set reconstruction based on spatially constrained multivariate clustering model

determined by reconstructed triangular networks may lead to deviations from original areal features. Certain units without actual spatial adjacency will be marked as neighboring objects, and such results fail to meet research expectations.

Spatial weight files constitute another widely adopted spatial constraint form. Research with simultaneous requirements on spatial proximity and temporal consistency can adopt spatial weight matrices to satisfy analytical demands. The research objects of this study focus on territorial space. Research scope covers continuous areal features. Analytical design focuses solely on spatial constraints and excludes temporal impacts in current stages. Territorial space development features highlight close socio-economic connections among areal features. Transportation network development strengthens interaction between units that share common nodes or boundaries. Comprehensive comparison of applicable methods supports the final selection of the shared edge corner constraint. This spatial constraint type serves as the core rule for territorial space clustering in this study.

Table 4.2 Spatial constraint type

Type	Explanation	Application scope
Shared edges only	Elements can be classified into the same cluster only when one element shares a common edge with another element in the cluster	Suitable for clustering analysis of contiguous areal features
Shared edges and corners	Elements can be classified into the same cluster only when one element shares a common edge or vertex with another element in the cluster	Suitable for clustering analysis of contiguous areal features
Trimmed Delaunay triangulation	This method defines neighborhood relationships by converting original features into Thiessen polygons based on Delaunay triangulation	Suitable for point features and areal features
Spatial weight file	Spatial weight matrix files are generated through spatial weight matrix construction tools	Suitable for research that requires both spatial proximity and temporal similarity

The selection of cluster numbers depends on the comprehensive context of specific research scenarios. Research groups with adequate prior knowledge can identify cluster numbers through the observation of data features and spatial distribution. Most academic studies lack unified and standardized criteria for the determination of cluster numbers. Analytical expectations focus on clustering outcomes that adequately reflect the similarity and heterogeneity of spatial elements. The optimal number of clusters can be finalized through parameter vacancy setting and repeated experimental verification. The Calinski-Harabasz index serves as a valid indicator to measure clustering effectiveness [15]. This index calculates the ratio of variance between clusters to variance within clusters. The fundamental motivation for applying the spatially constrained multivariate clustering model rests on the measurement set reconstruction of the DEA framework. The traditional DEA model studies highlight the demand for a large sample size of DMUs. Sufficient unit quantity avoids weak comparison of efficiency values and reduced model discrimination derived from inadequate samples. Continuous improvement of DEA theory reshapes the definition of production possibility sets. Non-classical DEA models break the constraint of the convex envelope of all DMUs points. A limited number of DMUs will not impair the discriminative capacity of improved model systems. Existing theories do not set rigid numerical standards for DMUs. An overly small sample scale impedes in-depth analysis and accurate quantitative analysis. This study integrates available prior knowledge with blank parameter configuration and experimental trials. Combined methods support the reasonable determination of optimal cluster quantities. Such comprehensive strategies maintain internal similarity of territorial space inside each cluster. They also secure distinct differentiation of territorial space features among different clusters. Final clustering schemes conform to the quantity specification of

DMUs in DEA modeling. Systematic combination of multiple principles guarantees scientific and reliable clustering results.

Analytical fields constitute the fundamental basis for distinguishing similarity and dissimilarity of territorial space. Priority is given to fields with ratio, interval, or ordinal attributes during field selection. Proper selection of such indicators ensures stable operation and accurate calculation of clustering models. Standardized calculation is conducted on all field values throughout model operation. Standardized processing makes full use of analytical indicators and supports objective evaluation of indicator performance in clustering analysis. Statistical indicators cover mean value, standard deviation, minimum value, maximum value, and coefficient of determination. Comprehensive statistical summary identifies core fields with strong capacity for cluster differentiation. Relevant results also provide solid reference for subsequent optimization and adjustment of analytical indicators. Two major factor categories are incorporated into the selection of analytical fields in this study. These categories dominate the actual condition of territorial space use. The first category refers to inherent geographical background factors. Relevant indicators contain digital elevation model (DEM), slope gradient, evaluation score of production living ecological space, environmental index, and landscape index. The second category covers multidimensional socio-economic development factors. Typical indicators include fixed asset investment, economic output, total retail sales of consumer goods, population scale, and infrastructure development status. Geographical background factors and socio-economic factors jointly form the indicator system of analytical fields. The integrated indicator system supports the completion of scientific spatial clustering.

4.2.3 Model Coupling Mechanism: Integration of Spatial Clustering Model and Generalized DEA Model

Section 4.2.2 elaborates on the mechanism and process of measurement set reconstruction. The core operation divides DMUs into clusters. This procedure serves as the fundamental premise for accurate efficiency measurement. This section further discusses the coupling and integration between DMU grouping and efficiency measurement. The research realizes the combination of spatially constrained multivariate clustering model and the generalized DEA model. The coupling mechanism is demonstrated and explained from two dimensions, as shown in Fig. 4.5.

First, the sequential data coupling mechanism. This mechanism defines a data transmission link between the two models. One model transmits the calculated data values to the other, and this process corresponds to value passing in high-level programming languages. This coupling mode presents simple logical correlation between models. It also ensures favorable independent performance and minimal mutual interference among different modules, and maintains a low overall coupling level. For these advantages, scholars widely apply such model coupling in relevant research. In this study, output results of the spatially constrained multivariate clustering model are imported into the generalized DEA model. The coupling of the two

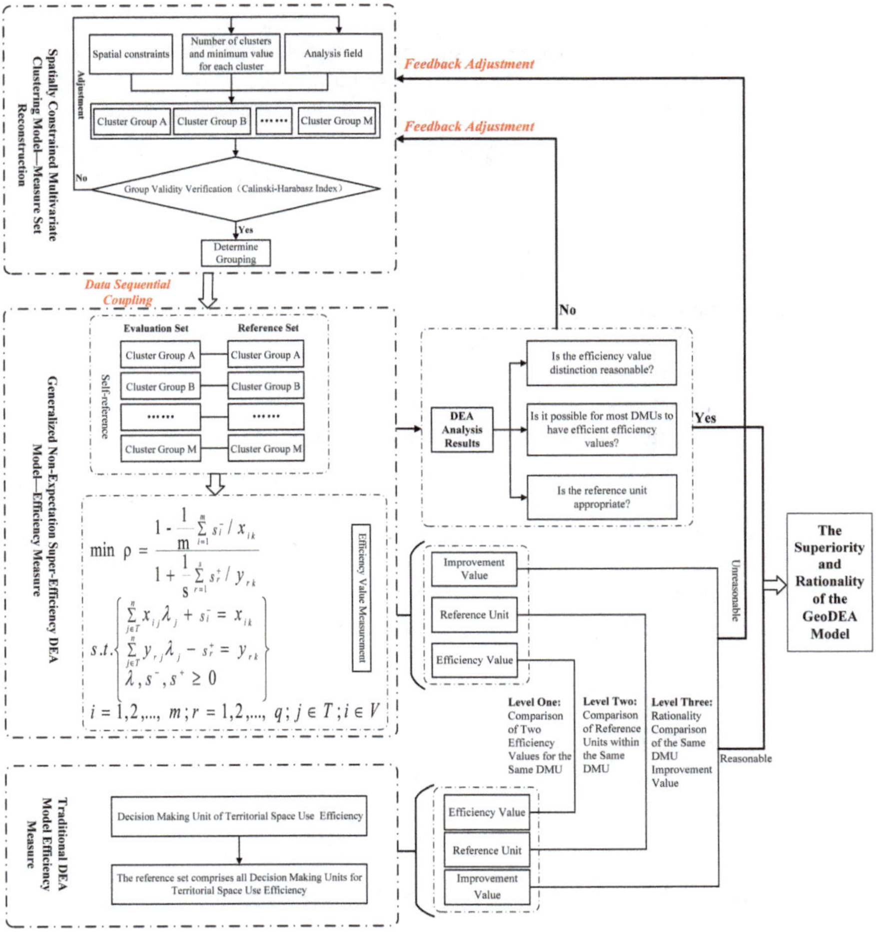

Fig. 4.5 Coupling mechanism between spatially constrained multivariate clustering model and generalized data envelopment analysis model

models further forms the GeoDEA model. Under the control of spatial constraints, clustering quantities and analysis indicators, the former model completes clustering and grouping of DMUs. It reconstructs measurement sets of TSUE with high similarity in geographical background and socio-economic conditions. The reconstructed measurement sets are subsequently delivered to the generalized DEA model. Data derived from the previous model provides basic inputs for the follow-up model. The generalized DEA model adopts reasonable group comparison modes and standardized parameter settings. On this basis, the research completes the quantitative measurement of TSUE.

Second, the feedback loop mechanism. This mechanism reflects a mutual feedback relationship of data outputs between two models. The output results of the former model can support the systematic evaluation of the feasibility and accuracy of the latter model. Targeted guidance can thus be provided for the data selection and method design of the latter model. Meanwhile, the outputs of the latter model can also verify the reliability of results generated by the former model. Corresponding adjustment and optimization can be conducted for the former model according to the verification conclusions. In this study, the practical value of the feedback loop mechanism is mainly reflected in the following contents. The efficiency measurement results of the generalized DEA model can help verify the rationality and reliability of the spatially constrained multivariate clustering model and offer clear guidance for subsequent model optimization. Relevant analysis covers three core dimensions including efficiency values, reference units, and improvement values. This study analyzes the efficiency results of the generalized DEA model based on reconstructed measurement sets and fully evaluates the rationality and accuracy of relevant measurement outcomes. Such analysis can further verify the rationality and effectiveness of measurement set reconstruction completed by the spatially constrained multivariate clustering model. If follow-up analysis shows that the output performance of the generalized DEA model does not conform to research requirements, researchers can adjust the parameter settings of the spatially constrained multivariate clustering model in a targeted manner. Such targeted modification can realize the dynamic adjustment and improvement of measurement set reconstruction.

4.3 Implementation and Verification of Geographic Data Envelopment Analysis: A Complete Methodological Framework

4.3.1 Design of Comparative Model Experiments

This study designs a systematic comparative experiment to verify the rationality and superiority of the GeoDEA model in measuring TSUE. It also aims to identify the essential differences between the GeoDEA model and the traditional DEA models. A controlled variable framework and unified criteria are applied to ensure the objectivity and comparability of experimental results. The experiment is implemented based on ArcGIS Pro 3.0 and PyCharm 2023.2.1. Python 3.10 is used for program writing and data processing. The GeoDEA model program is independently developed in this study, while the traditional DEA model adopts well-established algorithms to ensure the stability of the operation process and the reliability of results. The experimental design adheres to the principles of unified foundation, differential variables, and multi-dimensional comparison. Its core contents are as follows:

First, unification of samples and data. A total of 340 cities nationwide in 2020 are selected as DMUs, covering different scale types and four major regions to ensure the representativeness of the samples. All basic data are derived from statistical surveys and data from published literature, which are adopted for analysis after data cleaning, standardization, and outlier elimination. The measurement indicator system adopts the one constructed by the GeoFA model in Chap. 3. Its specific connotation and calculation method will be elaborated in detail in Chap. 6, and this chapter only directly uses this system. The total TSUE is obtained by weighted aggregation of production space use efficiency (PSUE), LSUE, and ecological space use efficiency (ESUE). The weights are determined by the entropy weight method, and the specific weight calculation process is detailed in Chap. 6.

Second, unification of parameters and environment. The two models adopt the same parameter combination, with differences only in the grouping of DMUs and the selection of reference sets. The traditional DEA models use a global reference set and do not consider the geospatial effect. The GeoDEA model completes the grouping of DMUs through spatially constrained multivariate clustering, and then constructs a local reference set. Both models operate based on the same toolbox, effectively eliminating the calculation errors caused by tool differences.

Third, experimental modules and comparison dimensions. The experiment is divided into two major modules: operation and result comparison. The operation module is responsible for running the two models and outputting core calculation results such as efficiency values and non-radial inefficiency values. The result comparison module designs comparison indicators around three dimensions, focusing on in-depth comparative analysis of 7 megacities and 14 very large cities. For small and medium-sized cities, only the consistency of the measurement trends of the two models is verified.

Fourth, experimental verification. The operational stability of the two models is verified through repeated runs. Combined with geospatial logic and the actual situation of urban development, a multi-dimensional inspection framework is designed to fully ensure the rationality and credibility of the conclusions of the comparative experiment.

4.3.2 Model Rationality Verification: A Multi-dimensional Inspection Framework

4.3.2.1 Principle Rationality: Consistency between Geographic Logic and Mathematical Formulation

In the context of rapid urban development, urban administrators seek high-quality benchmark cities to elevate overall development quality. Such benchmark cities share similar geographical conditions, sound socio-economic status, and coordinated development strategies. Administrators examine thoroughly advantageous industrial modes, characteristic development layouts, and ecological governance practices of typical benchmark cities. Relevant summary and gap analysis can offer reliable references for targeted urban upgrading, renewal, and sustainable development. The selection of benchmark cities does not merely focus on the most advanced cities across the country. The city of Rizhao in Shandong Province can serve as a typical example. As a coastal city of fourth-tier scale, Rizhao tends to select Weihai and Yantai within the same province as benchmark objects rather than Shanghai. Shanghai represents a leading megacity with unique development paths that do not align with local development conditions of Rizhao. Weihai and Yantai maintain high similarities with Rizhao in marine fishery, tourism industry, and policy implementation. The selection of adjacent and homogeneous cities can guarantee greater rationality for comparative analysis. By analogy, the application of a unified national reference frontier will impose prominent limitations on the evaluation of TSUE. A single set of reference units may lead to numerous biased evaluation results among research samples. Such limitation restricts in-depth analysis on TSUE and hinders targeted optimization and improvement. Moreover, the efficiency values calculated by the DEA model reflect relative evaluation results. Appropriate selection of reference frontiers therefore becomes an essential prerequisite for objective measurement. For the evaluation of TSUE, the theoretical logic of the generalized DEA model shows stronger applicability among various extended DEA methods. The traditional DEA models take only the most efficient DMUs as benchmark references. The evaluation of TSUE requires a wider range of comparison objects. Reasonable benchmark units should feature consistent geographical background, coordinated socio-economic progress, and equivalent development positioning, instead of focusing solely on top-ranked cities. This research follows the theoretical framework of the generalized DEA model and constructs the GeoDEA model on this basis.

A small number of existing studies have verified the rationality of coupling clustering analysis with the generalized DEA model. Relevant exploration exerts important influence on theoretical research of the generalized DEA model [13, 14]. Some scholars adopted the shortest distance method to calculate inter-group distance and complete sample clustering. The comprehensive measurement of per capita economic development across Chinese provinces and municipalities served as empirical evidence to prove the feasibility of combining clustering analysis and the generalized DEA model. Zhang and other researchers further discussed the integrated application of clustering analysis and the generalized DEA model. Their research focuses on accurate efficiency evaluation for DMUs with large quantitative gaps. Simple numerical examples were adopted to confirm the effectiveness of such combined methods [16]. Some scholars have integrated clustering analysis and the DEA model to evaluate learning efficiency of university graduates and quality assessment of building materials [17, 18]. In terms of land use efficiency assessment, Yu and colleagues carried out preliminary attempts [19]. Their research adopted the k-means algorithm to classify urban agglomerations before conducting efficiency evaluation. Ma et al. summarized mainstream methods for land use efficiency research [20]. Their research also outlined future research directions for the combined application of spatial clustering methods and the DEA model. The above research findings provide solid theoretical support for the coupling design of clustering analysis and the generalized DEA model in this study. Commonly adopted clustering methods in current literature remain relatively simple. Few studies have realized clustering classification from the perspective of geographic spatial characteristics. Territorial space presents obvious features of geographic similarity, spatial correlation, and regional heterogeneity. This study takes such unique attributes into full consideration. The spatially constrained multivariate clustering model is applied to realize spatial clustering of territorial space units.

To sum up, the coupling of the spatially constrained multivariate clustering model and the generalized DEA model in this research gains solid rationality from both realistic background and existing academic findings.

4.3.2.2 Result Rationality: Consistency Test with Realistic Cognition

This study measures the use efficiency of production space, living space, and ecological space, as well as the total TSUE of 340 cities across China in 2020. The empirical calculation verifies the rationality of the GeoDEA model. This research establishes an efficiency measurement indicator system based on the GeoFA model proposed in Chap. 3. All subsequent efficiency evaluation work proceeds within this unified framework. The total TSUE derives from the weighted summation of three types of spatial use efficiency. Weight coefficients are determined through the entropy weight method. The corresponding weights of production space, living space, and ecological space are 0.4431, 0.4272 and 0.1297, respectively. Relevant methodological details and calculation processes are elaborated in Chap. 6. Due to chapter layout and space constraints, this section focuses on quantitative and qualitative analysis of TSUE in megacities and very large cities. Cities of other scales present similar changing trends and research conclusions. Relevant analysis will not be repeated in this section to avoid repetitive discussion. This study refers to the national standard for urban scale classification issued by the State Council. The research also adopts official data from the Seventh National Population Census in 2020. Seven megacities and fourteen very large cities are finally selected as key research samples. The selected megacities include Shanghai, Beijing, Shenzhen, Chongqing, Guangzhou, Chengdu, and Tianjin. The selected very large cities cover Wuhan, Xi'an, Hangzhou, Foshan, Nanjing, Shenyang, Qingdao, Jinan, Changsha, Harbin, Zhengzhou, Kunming, Dalian, and Dongguan. Both the traditional DEA model and the GeoDEA model are applied in this study to complete the measurement of TSUE.

From the qualitative perspective, seven megacities rank at the national leading level in economic development, human settlement environment, technological progress, and ecological governance. These cities play key leading and exemplary roles and exemplary functions across the country. Outstanding performance in multiple dimensions indicates that such cities generally maintain medium and high levels of TSUE. From the quantitative perspective, Fig. 4.6 and Table 4.3 present intuitive calculation results. The traditional DEA model and GeoDEA model generate minor numerical gaps in the measurement of TSUE for Shanghai, Beijing, and Shenzhen. Efficiency values obtained from both models remain close to one and belong to the medium-to-high efficiency level. Such numerical findings conform to qualitative judgment and further prove the effectiveness and rationality of the GeoDEA model. Obvious numerical differences can be observed between two model results for specific evaluation items. Large gaps appear in LSUE of Guangzhou, Chengdu, and Tianjin, as well as PSUE of Chongqing. Relevant data features are presented in Fig. 4.6.

According to the statistical data in Table 4.3, the traditional DEA model yields a value of 0.1388 for the PSUE of Chongqing. Corresponding results for LSUE reach 0.0029 in Guangzhou, 0.0010 in Chengdu and 0.0002 in Tianjin. All above values reflect an extremely low-efficiency status. Such evaluation outcomes deviate greatly from objective understanding of urban development conditions. The GeoDEA model delivers more reasonable and credible measurement results under the constraint of local grouping rules. This model calculates the PSUE of Chongqing at 1.0457. Measured values of LSUE reach 1.0027 in Guangzhou, 1.0264 in Chengdu and 0.5892 in Tianjin. The above evaluation outcomes match practical cognition and actual development characteristics of sample cities.

In accordance with the analytical logic applied to megacities, fourteen very large cities also play prominent leading and driving roles in regional development. The TSUE of such cities should generally maintain a medium or medium-to-high level. Further quantitative analysis refers to Fig. 4.7 and Table 4.4. The numerical gaps between results from the traditional DEA model and the GeoDEA model are less than 0.45 for Foshan, Nanjing, Harbin, and Dongguan. The two models generate identical values for PSUE in Hangzhou and ESUE in Foshan. Such findings further verify the reliability and accuracy of the GeoDEA model in efficiency evaluation. Table 4.4 presents obvious defects in the calculation results of the traditional DEA model. The model produces extremely low values for LSUE in several sample cities. Relevant data include 0.0019 for Hangzhou, 0.0035 for Jinan, 0.0087 for Changsha, 0.0003 for Zhengzhou, and 0.0092 for Kunming. The model also records low ESUE of 0.1541 for Qingdao and 0.1465 for Dalian. All these values reflect an extremely inefficient state and contradict objective perceptions of urban development. The GeoDEA model provides more reasonable evaluation outcomes under optimized comparison frameworks. Measured LSUE reaches 0.6293 in Hangzhou, 1.1213 in Jinan, 1.0143 in Changsha, 1.1093 in Zhengzhou, and 1.0205 in Kunming. The ESUE of Qingdao and Dalian is calculated as 1.0793 and 1.0008, respectively. These evaluation outcomes conform closely to the actual development status and inherent characteristics of the above cities.

In summary, the traditional DEA models may underestimate real efficiency levels in the measurement of TSUE. This method may misclassify high-efficiency territorial units as low-efficiency ones. Such evaluation deviation fails to reflect the real development condition of territorial space. The biased results impede targeted optimization of production space, living space, and ecological space. These outcomes also restrict the realization of high-quality development goals for territorial space. The GeoDEA model reconstructs the measurement set by adopting spatially constrained multivariate clustering. This optimization improves the objectivity and accuracy of TSUE

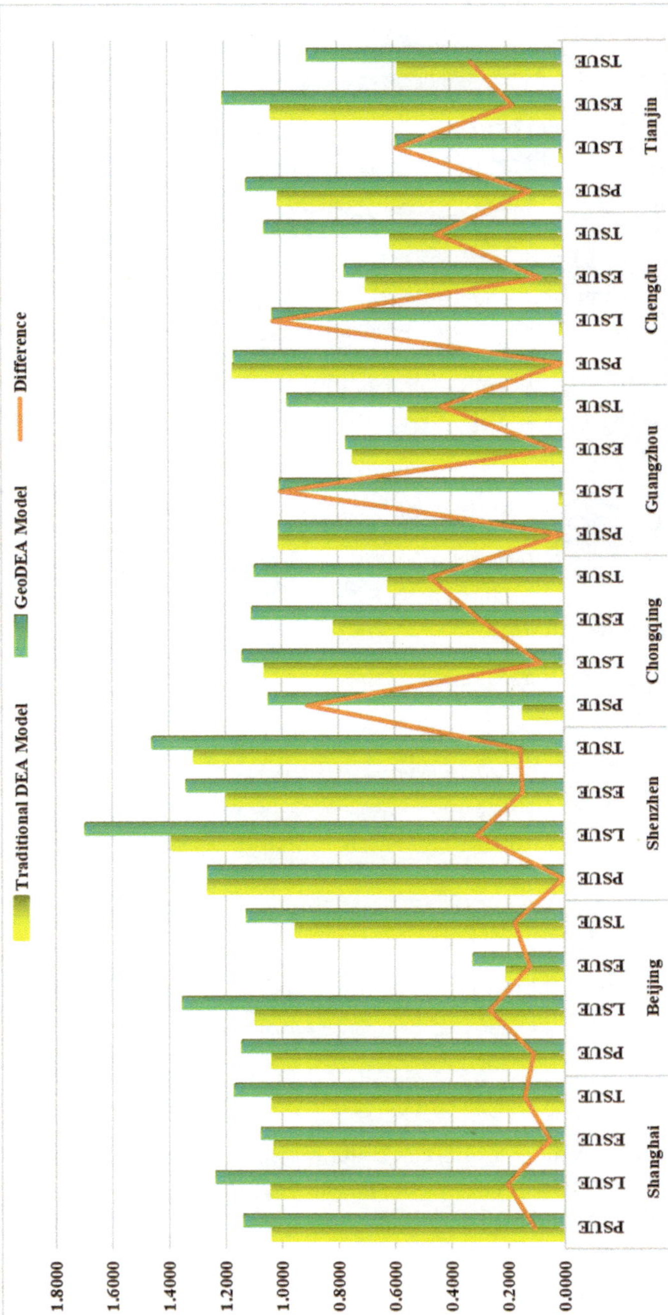

Fig. 4.6 Comparison of efficiency measurement results between traditional data envelopment analysis model and geographic data envelopment analysis model (Megacities)

Table 4.3 Comparison of territorial space use efficiency measurement results for seven megacities: traditional data envelopment analysis model and geographic data envelopment analysis model

City	Efficiency Type	Traditional DEA Model Result	GeoDEA Model Result	Difference
Shanghai	PSUE	1.0281	1.1357	0.1076
	LSUE	1.0310	1.2336	0.2026
	ESUE	1.0213	1.0736	0.0523
	TSUE	1.0284	1.1694	0.1410
Beijing	PSUE	1.0304	1.1396	0.1092
	LSUE	1.0872	1.3528	0.2656
	ESUE	0.2007	0.3228	0.1221
	TSUE	0.9471	1.1247	0.1776
Shenzhen	PSUE	1.2576	1.2620	0.0044
	LSUE	1.3850	1.6945	0.3095
	ESUE	1.1893	1.3375	0.1482
	TSUE	1.3032	1.4566	0.1534
Chongqing	PSUE	0.1388	1.0457	0.9069
	LSUE	1.0544	1.1359	0.0815
	ESUE	0.8067	1.1026	0.2959
	TSUE	0.6165	1.0916	0.4751
Guangzhou	PSUE	1.0031	1.0080	0.0049
	LSUE	0.0029	1.0027	0.9998
	ESUE	0.7397	0.7695	0.0298
	TSUE	0.5417	0.9748	0.4331
Chengdu	PSUE	1.1642	1.1645	0.0003
	LSUE	0.0010	1.0264	1.0254
	ESUE	0.6910	0.7726	0.0816
	TSUE	0.6059	1.0547	0.4488
Tianjin	PSUE	1.0012	1.1198	0.1186
	LSUE	0.0002	0.5892	0.5890
	ESUE	1.0254	1.2048	0.1794
	TSUE	0.5767	0.9041	0.3274

evaluation. Final measurement results can better match practical cognition and realistic urban features. The GeoDEA model presents higher precision and objectivity than the traditional DEA models. This new method delivers prominent practical value for the comprehensive measurement of TSUE.

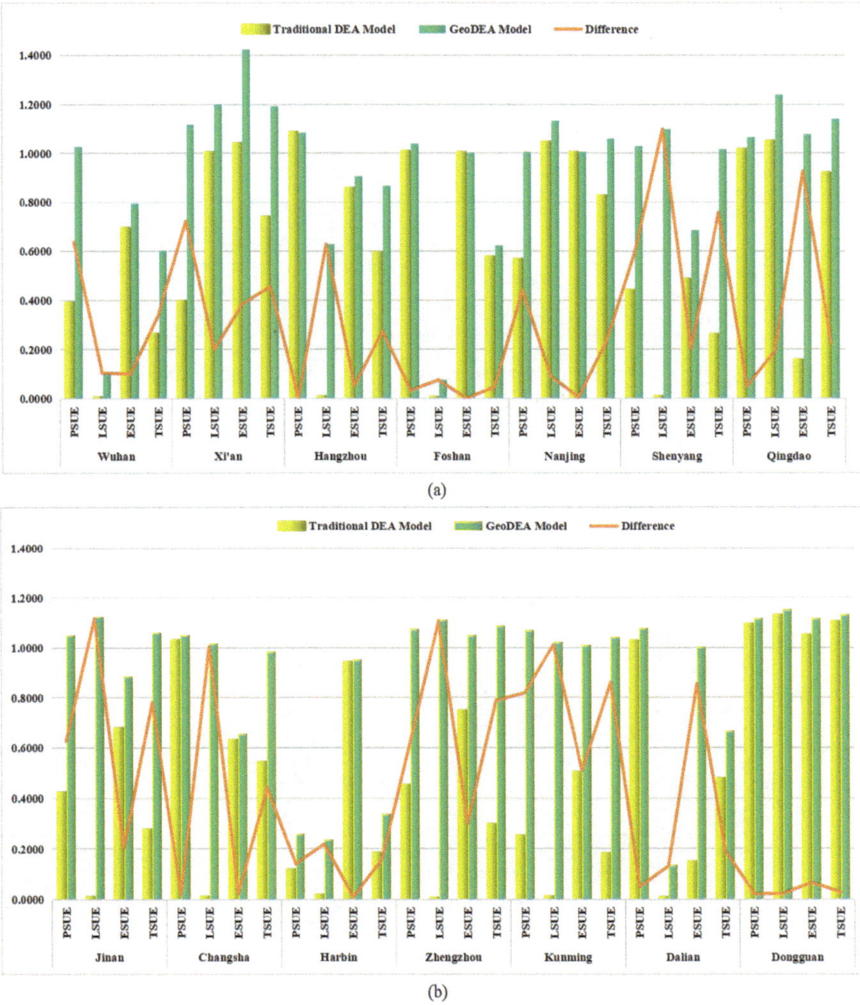

Fig. 4.7 Comparison of efficiency measurement results between traditional data envelopment analysis model and geographic data envelopment analysis model (very large cities)

4.3.2.3 Frontier Rationality: Geographic Interpretability of Spatially Constrained Frontiers

Following the research logic of Sect. 4.3.2.2, this section continues to focus on TSUE of megacities and very large cities. The discussion aims to further explore the rationality of reference frontiers derived from the GeoDEA model measurement results. This study selects typical evaluation items for targeted analysis. The selected items cover multiple types of spatial use efficiency across sample cities. The research takes LSUE of three megacities, including Guangzhou, Chengdu, and Tianjin as key

Table 4.4 Comparison of territorial space use efficiency measurement results for fourteen very large cities: traditional data envelopment analysis model and geographic data envelopment analysis model

City	Efficiency Type	Traditional DEA Model Result	GeoDEA Model Result	Difference
Wuhan	PSUE	0.3902	1.0272	0.6370
	LSUE	0.0006	0.1055	0.1049
	ESUE	0.6938	0.7950	0.1012
	TSUE	0.2631	0.6033	0.3402
Xi'an	PSUE	0.3961	1.1193	0.7232
	LSUE	1.0019	1.2013	0.1994
	ESUE	1.0377	1.4224	0.3847
	TSUE	0.7381	1.1936	0.4555
Hangzhou	PSUE	1.0843	1.0843	0.0000
	LSUE	0.0019	0.6293	0.6274
	ESUE	0.8553	0.9071	0.0518
	TSUE	0.5922	0.8669	0.2747
Foshan	PSUE	1.0060	1.0395	0.0335
	LSUE	0.0003	0.0759	0.0756
	ESUE	1.0024	1.0024	0.0000
	TSUE	0.5759	0.6230	0.0471
Nanjing	PSUE	0.5648	1.0061	0.4413
	LSUE	1.0436	1.1344	0.0908
	ESUE	1.0021	1.0066	0.0045
	TSUE	0.8261	1.0610	0.2349
Shenyang	PSUE	0.4395	1.0305	0.5910
	LSUE	0.0019	1.0993	1.0974
	ESUE	0.4831	0.6861	0.2030
	TSUE	0.2582	1.0152	0.7570
Qingdao	PSUE	1.0153	1.0651	0.0498
	LSUE	1.0476	1.2392	0.1916
	ESUE	0.1541	1.0793	0.9252
	TSUE	0.9174	1.1413	0.2239
Jinan	PSUE	0.4207	1.0476	0.6269
	LSUE	0.0035	1.1213	1.1178
	ESUE	0.6785	0.8821	0.2036
	TSUE	0.2759	1.0576	0.7817
Changsha	PSUE	1.0285	1.0491	0.0206
	LSUE	0.0087	1.0143	1.0056
	ESUE	0.6294	0.6533	0.0239

(continued)

Table 4.4 (continued)

City	Efficiency Type	Traditional DEA Model Result	GeoDEA Model Result	Difference
	TSUE	0.5411	0.9829	0.4418
Harbin	PSUE	0.1165	0.2557	0.1392
	LSUE	0.0155	0.2344	0.2189
	ESUE	0.9405	0.9499	0.0094
	TSUE	0.1802	0.3366	0.1564
Zhengzhou	PSUE	0.4494	1.0727	0.6233
	LSUE	0.0003	1.1093	1.1090
	ESUE	0.7456	1.0475	0.3019
	TSUE	0.2960	1.0851	0.7891
Kunming	PSUE	0.2498	1.0679	0.8181
	LSUE	0.0092	1.0205	1.0113
	ESUE	0.5006	1.0085	0.5079
	TSUE	0.1795	1.0400	0.8605
Dalian	PSUE	1.0266	1.0768	0.0502
	LSUE	0.0012	0.1317	0.1305
	ESUE	0.1465	1.0008	0.8543
	TSUE	0.4744	0.6632	0.1888
Dongguan	PSUE	1.0927	1.1142	0.0215
	LSUE	1.1277	1.1505	0.0228
	ESUE	1.0486	1.1136	0.0650
	TSUE	1.1019	1.1297	0.0278

objects. PSUE covers Wuhan, Xi'an, Nanjing, Shenyang, Jinan, and Zhengzhou. LSUE involves Hangzhou, Shenyang, Jinan, Changsha, Zhengzhou, and Kunming. ESUE includes Shenyang, Qingdao, Jinan, Zhengzhou, Kunming, and Dalian. Relevant measurement results serve as basic materials to demonstrate and analyze the rationality of reference frontiers. The subsequent analysis proceeds on the basis of efficiency data of the above cities.

As shown in Table 4.5, the traditional DEA model obtains low values of LSUE for Guangzhou, Chengdu, and Tianjin. The reference frontier of all three cities includes Bortala Mongolian Autonomous Prefecture. The selection of this prefecture as the benchmark unit is unreasonable and offers no practical guiding value. First, there are substantial geographical distances between the three megacities and Bortala Mongolian Autonomous Prefecture. The two sides present great differences in landform, terrain, and climatic conditions. Second, the three megacities feature large population size, developed economy, and strong scientific and technological innovation capacity. Such advantages promote rapid socio-economic development in urban areas. Bortala Mongolian Autonomous Prefecture has a much smaller population scale. Local economic growth and urban construction develop at a relatively slow rate. Third, the non-radial inefficiency values calculated by the traditional DEA model remain at a high level. Such results deviate from basic common sense and actual urban development conditions, especially with consideration of local resource input scale. The GeoDEA model provides more reasonable measurement outcomes with optimized reference combinations. The reference frontier of LSUE in Guangzhou consists of Dongguan, Quanzhou, and Shenzhen. The corresponding frontier of Chengdu includes Dazhou, Karamay, and Chongqing. The reference units for Tianjin cover Beijing, Cangzhou, and Tangshan. Cities on such improved frontiers share close geographical locations with research samples. These units also maintain similar economic development modes and development stages. The grouping mode ensures higher rationality for comparative evaluation. Moreover, the GeoDEA model generates low non-radial inefficiency values for the three megacities. Megacities occupy high administrative levels and lead national development in multiple fields. Mature industrial structures and complete economic systems have taken shape in these regions. Small non-radial inefficiency values conform fully to the inherent development characteristics of high-level megacities.

This section analyzes the rationality of reference frontiers for very large cities. The research conducts detailed discussion on cities with significant disparities between the results of two models, as illustrated in Tables 4.6, 4.7 and 4.8. In terms of output non-radial inefficiency, the traditional DEA model and the GeoDEA model

Table 4.5 Comparison of reference frontiers for efficiency measurement of the same decision-making unit under two models (LSUE of megacities)

DMUs	DEA model	LSUE	Output non-radial inefficiency	Reference frontier
Guangzhou	Traditional DEA model	0.0029	339.0343	Bortala Mongolian autonomous prefecture (0.0594); Quanzhou (0.4126); Shanghai (0.1557); Shaoxing (0.0010); Shenzhen (0.4344); Chongqing (0.0689)
	GeoDEA model	1.0027	−0.0027	Dongguan (0.3918); Quanzhou (0.4517); Shenzhen (0.6019)
Chengdu	Traditional DEA model	0.0010	1028.4464	Bortala Mongolian autonomous prefecture (0.3731); Nanjing (0.2110); Shanghai (0.1108); Chongqing (0.4084)
	GeoDEA model	1.0264	−0.0258	Dazhou (0.4715); Karamay (0.1497); Chongqing (0.5200)
Tianjin	Traditional DEA model	0.0002	4495.3649	Bortala Mongolian autonomous prefecture (0.8501); Shanghai (0.1832); Shaoxing (0.0329); Shenzhen (0.0626)
	GeoDEA model	0.5892	0.6972	Beijing (0.1947); Cangzhou (0.8369); Tangshan (0.2896)

show minor differences in the evaluation of PSUE and ESUE. Noticeable divergence emerges in the measurement of LSUE. The traditional DEA model generates high output non-radial inefficiency values. The GeoDEA model obtains much lower values for the same indicator. High inefficiency values usually indicate excessive resource input in very large cities. Excessive resource allocation further triggers severe resource waste. Such evaluation findings fail to fit the actual development stage of very large cities. The GeoDEA model thus possesses higher rationality than the traditional DEA model in practical efficiency measurement.

Table 4.6 Comparison of reference frontiers for efficiency measurement of the same decision-making unit under two models (PSUE of very large cities)

DMUs	DEA model	PSUE	Output non-radial inefficiency	Reference frontier
Wuhan	Traditional DEA model	0.3902	1.5626	Chengdu (0.3200); Luohe (0.6121); Qingdao (0.1127); Shenzhen (0.2511); Zhoushan (0.4215)
	GeoDEA model	1.0272	−0.0265	Hefei (0.2457); Shanghai (0.1542); Wuxi (0.0678); Xuzhou (0.0539); Zhenjiang (0.8211)
Xi'an	Traditional DEA model	0.3961	1.5245	Luohe (1.6175); Shenzhen (0.3144); Zhoushan (0.3554)
	GeoDEA model	1.1193	−0.1066	Beijing (0.1674); Guyuan (0.5742); Karamay (0.2812); Panjin (0.0064)
Nanjing	Traditional DEA model	0.5648	0.7705	Changsha (0.2059); Qingdao (0.1459); Shenzhen (0.1860); Suzhou (0.0551); Wuxi (0.2518); Xuchang (0.2160); Zhoushan (0.2602)
	GeoDEA model	1.0061	−0.0061	Jiaxing (0.7459); Shanghai (0.0547); Suzhou (0.2415); Wuxi (0.3007)
Shenyang	Traditional DEA model	0.4395	1.2756	Beijing (0.0217); Guyuan (0.0693); Hengshui (0.4032); Shanghai (0.0936); Xingtai (0.6373); Zhoushan (0.0738)
	GeoDEA model	1.0305	−0.0296	Beijing (0.0586); Chengdu (0.0524); Dalian (0.2832); Panjin (0.1554); Qingyang (0.2217)
Jinan	Traditional DEA model	0.4207	1.3768	Bozhou (0.6960); Chengdu (0.1001); Jiayuguan (0.3269); Shenzhen (0.2534); Tiemenguan (0.0981)
	GeoDEA model	1.0476	−0.0455	Bozhou (0.1706); Huaibei (0.0507); Rizhao (0.6850); Zhengzhou (0.4695)
Zhengzhou	Traditional DEA model	0.4494	1.2253	Bozhou (0.5785); Chengdu (0.0201); Shenzhen (0.3615); Zhoushan (0.3980)
	GeoDEA model	1.0727	−0.0678	Jinan (0.3433); Jiyuan (0.4314); Qingdao (0.2692)

Table 4.7 Comparison of reference frontiers in efficiency measurement for the same decision-making unit under two models (LSUE of very large cities)

DMUs	DEA model	LSUE	Output non-radial inefficiency	Reference frontier
Hangzhou	Traditional DEA model	0.0019	536.3482	Beijing (0.0914); Bortala Mongolian autonomous prefecture (0.5722); Dongguan (0.3377); Nanjing (0.0597); Shanghai (0.1801); Chongqing (0.0178)
	GeoDEA model	0.6293	0.5892	Quanzhou (0.6848); Shaoxing (0.1504); Shenzhen (0.2239); Shennongjia Forestry District (0.0790); Xiamen (0.0335)
Shenyang	Traditional DEA model	0.0019	519.2843	Bortala Mongolian autonomous prefecture (0.2161); Qingdao (0.1279); Quanzhou (0.5475); Shaoxing (0.0337); Zunyi (0.0189)
	GeoDEA model	1.0993	−0.0903	Baotou (1.0983); Xi'an (0.0002); Zhengzhou (0.2606)
Jinan	Traditional DEA model	0.0035	280.8307	Beijing (0.0016); Bortala Mongolian autonomous prefecture (0.0881); Qingdao (0.6862); Quanzhou (0.1270); Shaoxing (0.0412); Urumqi (0.1072)
	GeoDEA model	1.1213	−0.1082	Beijing (0.0689); Langfang (0.7631); Tangshan (0.3460)
Changsha	Traditional DEA model	0.0087	113.3318	Bortala Mongolian autonomous prefecture (0.0999); Kunyu (0.1975); Qingdao (0.5303); Quanzhou (0.0452); Shaoxing (0.0103); Shenzhen (0.0247); Chongqing (0.1043)
	GeoDEA model	1.0143	−0.0141	Dongguan (0.0648); Quanzhou (0.6080); Shenzhen (0.1073); Shennongjia Forestry District (0.0081); Zhuhai (0.1172)
Zhengzhou	Traditional DEA model	0.0003	2956.8914	Beijing (0.0880); Bortala Mongolian autonomous prefecture (0.4072); Qingdao (0.0003); Quanzhou (0.3798); Shanghai (0.0089); Chongqing (0.1458)
	GeoDEA model	1.1093	−0.0985	Shenyang (0.6720); Taiyuan (0.6571); Xi'an (0.1437)

(continued)

Table 4.7 (continued)

DMUs	DEA model	LSUE	Output non-radial inefficiency	Reference frontier
Kunming	Traditional DEA model	0.0092	108.1381	Bortala Mongolian autonomous prefecture (0.2177); Quanzhou (0.5403); Zhanjiang (0.0585); Zunyi (0.1326)
	GeoDEA model	1.0205	−0.0201	Dazhou (0.1653); Karamay (0.1932); Qujing (0.2901); Yinchuan (0.1349); Chongqing (0.1672); Zunyi (0.0905)

In terms of reference frontiers, the frontier units identified by the GeoDEA model share higher similarity in geographical location, environmental conditions, social development, and economic progress. This section selects the evaluation results of Wuhan to elaborate on the comparison of PSUE, as presented in Table 4.6. The reference frontier of the traditional DEA model consists of Chengdu, Luohe, Qingdao, Shenzhen, and Zhoushan. Wuhan functions as a core heavy industry base in China. The city also develops high-tech and emerging industries. Wuhan is a very large city that combines traditional industrial sectors and emerging industrial systems. Chengdu takes modern service industry, manufacturing, and agriculture as its major economic pillars. Luohe focuses on the development of light industry. Qingdao gains economic momentum from private business and export trade. Shenzhen features an export-oriented economic structure. Zhoushan relies on fishery, port economy, and tourism as core supporting industries. All benchmark cities in this frontier show prominent differences from Wuhan in development patterns and economic structures. The reference frontier of the GeoDEA model includes Hefei, Shanghai, Wuxi, Xuzhou, and Zhenjiang. From the perspective of geographical distribution, most cities in this group belong to the Yangtze River Economic Belt with advanced development levels. These cities present high developmental consistency with Wuhan. Xuzhou acts as the central city of the Huaihai Economic Circle. The local economic system is supported by both heavy industry and light industry. The city shares comparable economic features and regional strategic status with Wuhan. The reference frontier adopted by the GeoDEA model reflects stronger rationality and validity in measuring the PSUE of Wuhan.

In terms of LSUE, this section conducts a detailed analysis of the measurement results of Jinan as shown in Table 4.7. The reference frontier of the traditional DEA model consists of Beijing, Bortala Mongolian Autonomous Prefecture, Qingdao, Quanzhou, Shaoxing, and Urumqi. Bortala Mongolian Autonomous Prefecture and Urumqi feature a temperate continental arid climate. Quanzhou has a subtropical oceanic monsoon climate, and Shaoxing has a subtropical monsoon climate. Jinan belongs to the temperate monsoon climate zone. Climatic differences lead to distinct patterns in residents' living habits and daily activities across regions. Obvious discrepancies in climatic conditions weaken the rationality for Jinan to benchmark

Table 4.8 Comparison of reference frontiers for efficiency measurement of the same decision-making unit under two models (ESUE of very large cities)

DMUs	DEA model	ESUE	Output non-radial inefficiency	Reference frontier
Shenyang	Traditional DEA model	0.4831	1.0698	Baise (0.0843); Bortala Mongolian autonomous prefecture (0.0001); Hegang (0.2089); Huai'an (0.1640); Huaibei (0.3931); Yangzhou (0.0876)
	GeoDEA model	0.6861	0.4574	Hegang (0.2800); Jinzhou (0.0161); Panjin (0.1434); Tianjin (0.3480)
Qingdao	Traditional DEA model	0.1541	5.4904	Bengbu (0.0388); Bortala Mongolian autonomous prefecture (0.0004); Huaibei (0.4777); Jining (0.3177); Nanyang (0.0427); Suihua (0.0262)
	GeoDEA model	1.0793	−0.0735	Dongying (0.9079); Jining (0.0264); Nanyang (0.0211)
Jinan	Traditional DEA model	0.6785	0.4739	Baise (0.1114); Bortala Mongolian autonomous prefecture (0.0004); Jining (0.1506); Suzhou (0.4477); Zhoukou (0.1614)
	GeoDEA model	0.8821	0.1337	Dongying (0.4917); Huaibei (0.0157); Jining (0.2491); Jiyuan (0.2068)
Zhengzhou	Traditional DEA model	0.7456	0.3411	Bortala Mongolian autonomous prefecture (0.0000); Jining (0.2117); Jiaxing (0.6596); Shanghai (0.1430); Suzhou (0.0045)
	GeoDEA model	1.0475	−0.0454	Heze (0.3248); Jining (0.1413); Jiyuan (0.5292)
Kunming	Traditional DEA model	0.5006	0.9977	Baise (0.0964); Fangchenggang (0.4820); Jixi (0.4236); Mudanjiang (0.0018); Nujiang Lisu autonomous prefecture (0.0147); Siping (0.2415); Xining (0.0527)
	GeoDEA model	1.0085	−0.0085	Lijiang (0.0032); Nujiang Lisu autonomous prefecture (0.1922); Panzhihua (0.1627); Qujing (0.7087)

(continued)

Table 4.8 (continued)

DMUs	DEA model	ESUE	Output non-radial inefficiency	Reference frontier
Dalian	Traditional DEA model	0.1465	5.8249	Baise (0.0599); Bortala Mongolian autonomous prefecture (0.0001); Huaibei (0.1077); Mianyang (0.1503); Shuangyashan (0.1507); Siping (0.0023); Yangzhou (0.5654)
	GeoDEA model	1.0008	−0.0008	Hegang (0.2200); Hulunbuir (0.0061); Jinzhou (0.4933); Panjin (0.0308); Qinhuangdao (0.1104); Tianjin (0.2346)

against the above cities. The reference frontier determined by the GeoDEA model includes Beijing, Langfang, and Tangshan. All three cities share the same temperate monsoon climate with Jinan and maintain close geographical proximity. Residents in these regions present highly consistent living patterns and lifestyle characteristics. The reference frontier constructed by the GeoDEA model therefore delivers higher scientificity and rationality in efficiency evaluation.

In terms of ESUE, this section elaborates on the measurement results of Shenyang, as presented in Table 4.8. The reference frontier of the traditional DEA model includes Baise, Bortala Mongolian Autonomous Prefecture, Hegang, Huai'an, Huaibei, and Yangzhou. Bortala Mongolian Autonomous Prefecture is located in northwest China. Huai'an, Huaibei, and Yangzhou belong to southern regions. These cities differ greatly from Shenyang in ecological background and climatic conditions. Significant geographical distances also exist between such benchmark units and Shenyang. The reference frontier defined by the GeoDEA model consists of Hegang, Jinzhou, Panjin, and Tianjin. Except for Tianjin, the other three cities are all located in Liaoning Province together with Shenyang. These regions share highly similar ecological endowments and climatic features, which provides solid reference value for comparative evaluation. The reference frontier constructed by the GeoDEA model therefore demonstrates higher rationality and scientificity in efficiency measurement.

In summary, from the perspective of reference frontier rationality, the frontiers adopted by the GeoDEA model for efficiency measurement are generally reasonable and scientifically sound. Accordingly, the TSUE values calculated by the GeoDEA model are more accurate and persuasive.

Overall, Section 4.3.2 conducts a comprehensive and in-depth verification and analysis of the GeoDEA model and elaborates on the discrepancies in efficiency measurement results between the GeoDEA model and the traditional DEA model through detailed comparison. The core factor leading to inconsistent measurement results lies in the mathematical differences between the reference set and the measurement set. As DEA-series models measure relative efficiency, the selection of an appropriate reference set is critically important. Relying on the spatially constrained multivariate clustering model, this study realizes the reconstruction of the measurement set and further guarantees the reliability of TSUE results derived from the GeoDEA model.

4.3.3 Comprehensive Conclusions

Based on the verification and analysis in Sect. 4.3.1 and Sect. 4.3.2, this study summarizes key conclusions by comparing measurement differences and theoretical adaptability between two models. The analysis also considers the rationality of reference frontiers. Relevant core findings are presented as follows.

First, the construction of the GeoDEA model is scientifically sound and practically necessary. Territorial spatial units present obvious geographical heterogeneity and spatial dependence. The assumption of a globally unified reference set in the traditional DEA models fails to adapt to real evaluation requirements. The GeoDEA model adopts a coupling framework combining spatially constrained multivariate clustering and the generalized DEA model. The framework incorporates geospatial effects into the pre-evaluation process. This design realizes the grouping principle of intra-group homogeneity and inter-group disparity among DMUs. The logical system of the GeoDEA model conforms to geographical laws. Solid mathematical foundations and existing research findings further support its theoretical rationality.

Second, the GeoDEA model delivers higher measurement accuracy than the traditional DEA models. This research compares the evaluation results of 340 prefecture-level cities across China in 2020. The traditional DEA model substantially underestimates overall efficiency levels. The model misjudges partial TSUE of megacities such as Guangzhou and Chengdu as extremely inefficient. Similar evaluation biases also exist among very large cities, including Hangzhou and Jinan. Measurement outputs of the GeoDEA model match the actual urban development status more closely. The revised results correct systematic deviations and improve the objectivity and reliability of efficiency evaluation.

Third, the reference frontier of the GeoDEA model possesses stronger geographical interpretability. Reference units selected by the traditional DEA model lack close geographical connection. These benchmark cities also show marked differences in development models. Typical unreasonable cases include the selection of Bortala Mongolian Autonomous Prefecture as the benchmark for megacities. Such reference combinations provide limited practical guidance. In contrast, benchmark units of the GeoDEA model feature geographical proximity and coordinated development stages. These cities share consistent geographical endowments. The optimized reference frontier offers targeted benchmarks for future efficiency improvement.

Fourth, repeated tests fully confirm the stability and practicality of the GeoDEA model. Comparative experiments and multi-dimensional rationality examinations are conducted under unified experimental conditions. The GeoDEA model inherits the core strengths of classic DEA methods. These advantages include non-parametric estimation and flexible measurement for multiple inputs and multiple outputs. The introduction of geographical spatial characteristics addresses the major defect of the traditional DEA models. Optimized evaluation outcomes are highly consistent with actual urban development.

4.4 Chapter Summary

4.4.1 Core Characteristics and Methodological Contributions of GeoDEA Model

Centering on the core demand of TSUE measurement, this chapter breaks the inherent assumption of independent and identically distributed DMUs in the traditional DEA models. The study constructs and verifies the GeoDEA model with integrated geospatial effects. The core characteristics as well as academic and practical contributions of this model can be summarized into three aspects. In terms of model characteristics, the GeoDEA model forms an integrated framework covering spatial constraint, clustering grouping, and efficiency evaluation. The framework achieves three major breakthroughs. First, the model takes geospatial effects as the precondition of measurement. It breaks the limitations of the globally unified frontier adopted by traditional DEA models. Spatially constrained multivariate clustering realizes intra-group homogeneity and inter-group differentiation of DMUs.

Second, the model reconstructs local efficiency frontiers. Efficiency measurement shifts from global comparison to localized adaptation. This improvement fully accommodates regional differences in geographical endowments and socio-economic development conditions. Third, the model realizes structural coupling between spatial clustering and the generalized DEA model. It maintains the inherent advantages of the traditional DEA models including non-parametric estimation and multi-input and multi-output evaluation. The integration of spatial correlation also endows measurement results with mathematical rationality and geographical interpretability.

In terms of academic contributions, the GeoDEA model addresses the research gap of the traditional DEA models. Conventional efficiency evaluation often ignores geospatial effects in the measurement of TSUE. The new model enriches the theoretical framework of spatial DEA models and proposes a combined framework of spatially constrained multivariate clustering and generalized DEA model. This innovative framework provides a new methodological reference for efficiency evaluation targeting geographically correlated DMUs. Meanwhile, multi-dimensional verification clarifies the influence of mathematical relationships between reference sets and measurement sets on evaluation outputs. Relevant conclusions enhance the understanding of the essential attribute of relative efficiency in the DEA models theoretical system. In terms of practical contributions, the GeoDEA model addresses prominent defects of evaluation based on the traditional DEA models. Conventional measurement results often deviate from realistic conditions and adopt unreasonable reference frontiers. Evaluation outputs of the GeoDEA model reflect the actual level of territorial space use more accurately. The model delivers precise quantitative support for urban territorial space optimization and regional development planning. Relevant research outcomes further promote the realization of high-quality development goals in territorial spatial governance.

4.4.2 Applicable Scenarios and Application Notes of GeoDEA Model

The GeoDEA model presents clear pertinence in applicable scenarios. This model suits the efficiency measurement of multi-input and multi-output DMUs with obvious geographical spatial heterogeneity and spatial dependence. It shows high adaptability to research fields closely related to geographical location, including territorial space, regional economy, resources, and environment. Specifically, research objects often have obvious geographical differentiation features. Typical manifestations cover differences in terrain, climate, and resource endowment. Close spatial correlation also exists among DMUs in the forms of industrial radiation and resource flow. The traditional DEA models tend to produce prominent deviations under such conditions. Common problems include the underestimation of regional actual efficiency and the unreasonable geographical setting of reference frontiers. The application of the GeoDEA model can greatly improve the accuracy of relevant measurement results. The GeoDEA model demonstrates outstanding advantages in multiple typical situations. Representative cases contain the evaluation of urban TSUE at national or regional scales and the assessment of resource allocation efficiency among cities within urban agglomerations.

Three key points require adequate attention in the model application process. First, the selection of spatial constraint types should match the basic features of research objects. Territorial spatial units with continuous areal attributes need to give priority to applicable constraint rules suitable for polygon clustering such as adjacent edge and corner rules. Appropriate selection prevents clustering outcomes from deviating from objective geographical laws due to improper constraint design.

Second, the determination of cluster quantity and analysis indicators balances scientific rigor and practical operability. The number of clusters can be defined comprehensively through prior knowledge, experimental verification, and Calinski-Harabasz index. Core analysis indicators contain geographical background conditions and socio-economic factors simultaneously. Such indicator selection ensures high intra-group homogeneity. Third, the configuration of model parameters conforms to specific research objectives. The selection of input orientation or output orientation follows the actual operation status of territorial space use. Relevant settings also include the hypothesis of returns to scale and reference benchmark modes. Full consideration of undesirable outputs helps reduce systematic errors. Standardized parameter configuration stabilizes the overall reliability of final measurement results.

4.4.3 Limitations and Future Research Prospects of GeoDEA Model

The rationality and practicality of the GeoDEA model have been verified through multi-dimensional tests. Limited by research conditions and technical constraints, this model still has certain limitations. First, the spatially constrained multivariate clustering in model construction only takes geographical proximity and attribute similarity into account. The clustering framework lacks full integration of implicit spatial correlation factors. Relevant factors include traffic accessibility and policy relevance. Clustering results may fail to fully reflect the actual correlation characteristics of DMUs. Second, the super-efficiency SBM model adopted for efficiency measurement can address undesirable outputs. This method ignores the dynamic changes of input and output indicators. The static measurement framework cannot reflect the temporal evolution characteristics of TSUE. Third, existing model verification mainly focuses on megacities and very large cities. Sufficient tests are still absent for the applicability of this model in small and medium-sized cities and county-scale research. The general applicability of the GeoDEA model requires further examination and improvement.

Corresponding improvements and expansions can be carried out in four directions on the basis of the above limitations. First, future research can optimize the current spatial clustering method. The integration of implicit spatial factors such as traffic networks and policy zoning helps build a multi-dimensional spatially constrained clustering system. The revised system can improve the scientificity of grouping results for DMUs. Second, the introduction of dynamic DEA ideas and panel data supports the construction of a dynamic GeoDEA model. The improved model can realize temporal evolution analysis of TSUE. Relevant research can further identify the driving mechanisms behind efficiency changes. Third, subsequent studies can expand the application scale of the model. Extended evaluation covers TSUE in small and medium-sized cities and county units. Continuous inspection and parameter optimization can enhance the general applicability of the model in diverse research contexts. Fourth, in-depth coupling application can be realized between the GeoDEA model and spatial econometric models. Combined methods can explore the spatial correlation characteristics and influencing factors of TSUE. The research outcomes can provide comprehensive theoretical support and practical guidance for territorial spatial optimization.

Competing Interests The authors have no conflicts of interest to declare that are relevant to the content of this chapter.

References

1. Cook WD, Seiford LM (2009) Data envelopment analysis (DEA) - thirty years on. Eur J Oper Res 192(1):1–17
2. Kao C (2014) Network data envelopment analysis: a review. Eur J Oper Res 239(1):1–16
3. Zhou HB, Yang Y, Chen Y et al (2018) Data envelopment analysis application in sustainability: the origins, development and future directions. Eur J Oper Res 264(1):1–16
4. Banker RD, Charnes A, Cooper WW (1984) Some models for estimating technical and scale inefficiencies in data envelopment analysis. Manag Sci 30(9):1078–1092
5. Wei QL, Yan H (2010) A data envelopment analysis (DEA) evaluation method based on sample decision making units. Int J Inf Technol & Decis Mak 9(4):601–624
6. Zhou P, Ang BW, Poh KL (2008) A survey of data envelopment analysis in energy and environmental studies. Eur J Oper Res 189(1):1–18
7. Wu J, Liang L (2012) A DEA model for identifying critical input-output performance measures. J Syst Sci Complex 25(2):275–286
8. Yan YQ, Tao J (2016) A study of the operation efficiency and cost performance indices of power-supply companies in China based on a dynamic network slacks-based measure model. Omega-Int J Manag Sci 60:85–97
9. Foresman T, Luscombe R (2017) The second law of geography for a spatially enabled economy. Int J Digit Earth 10(10):979–995
10. Goodchild MF (2004) The validity and usefulness of laws in geographic information science and geography. Ann Assoc Am Geogr 94(4):709–714
11. Walker RT (2022) Geography, VON Thunen, and Tobler's first law: tracing the evolution of a concept. Geogr Rev 112(4):591–607
12. Westlund H (2013) A brief history of time, space, and growth: Waldo Tobler's first law of geography revisited. Ann Reg Sci 51(3):917–924
13. Ma S J (2015) Generalized data envelopment analysis model evaluated based on sample unit and its properties. Dissertation, Inner Mongolia University
14. Ma SJ, Liu J, Ma ZX et al (2012) Application of cluster analysis to determine sample unit set in generalized DEA method. Math Pract Theory 42(12):28–36
15. Kozak M (2012) "a dendrite method for cluster analysis" by Calinski and Harabasz: a classical work that is far too often incorrectly cited. Commun Stat-Theory Methods 41(12):2279–2280
16. Zhang X Y (2023) Research on theory and application of generalized data envelopment analysis with uncertain factors. Dissertation, Inner Mongolia Agricultural University
17. Zeng L C (2020) Study on efficiency evaluation of college graduates based on DEA fusion algorithm. Dissertation, Jingdezhen Ceramic University
18. Xu Wei (2020) Building materials evaluation model based on cluster and interval DEA. Dissertation, Jingdezhen Ceramic University
19. Yu JQ, Zhou KL, Yang SL (2019) Land use efficiency and influencing factors of urban agglomerations in China. Land Use Policy 88:104143
20. Ma Y, Zheng MR, Zheng XQ et al (2023) Land use efficiency assessment under sustainable development goals: a systematic review. Land 12(4):894

Chapter 5
Trend Prediction: Construction of Extended Cellular Automata Model

Abstract This chapter focuses on the prediction of TSUE and develops the extended CA model. It first identifies the core limitations of the traditional CA models, including excessive reliance on empirical rules and insufficient capacity for geographic feature representation. On this basis, machine learning algorithms (CNN, LSTM, and RF) are incorporated to optimize the traditional CA models, forming a coupled CNN-LSTM-RF-CA framework. The proposed model integrates data-driven mining and knowledge-guided logic, extracts complex spatiotemporal features, and calculates cellular transition probabilities to significantly improve simulation accuracy. Empirical verification based on panel data of 340 Chinese cities demonstrates that the extended CA model outperforms the traditional model in multiple accuracy metrics. This study provides a reliable methodological approach for the dynamic prediction of TSUE.

5.1 From Traditional Cellular Automata Model to Extended Cellular Automata Model: The Evolution of Simulation Paradigm

5.1.1 Traditional CA Model: Basic Principles and Limitations in Simulation

The traditional CA models represent dynamic simulation models based on discrete space and discrete time [1]. The core mechanism of the traditional CA models lies in the iterative operation of simple local rules. Such operation can generate the global evolution behavior of complex systems. The basic principles of the traditional CA models revolve around four core elements including cell, space, state, and rule [2]. These elements connect and interact with one another. They form a complete logical system for the dynamic evolution of the model. Cells act as the fundamental spatial units of simulation and construct the entire cellular space. Cell states refer to specific attributes at a given moment with discrete and finite values. Neighborhood rules serve

Y. Ma, *Evaluation and Prediction Modeling of Territorial Space Use Efficiency*, Urban Sustainability, https://doi.org/10.1007/978-981-92-4396-9_5

as the core driver of model evolution. These rules define the interaction between each cell and its adjacent units. They further determine the probability and conditions of cell state transition and realize the dynamic simulation of complex system evolution.

The traditional CA models contain theoretical potential for the simulation of complex geographical phenomena. Even so, they exhibit structural defects in the expression of geographical features. Such shortcomings restrict the simulation performance for practical scenarios. Five major limitations can be summarized as follows. First, traditional CA models divide cellular space with regular geometric units. Such settings fail to match the irregular form of real geographical space. The model cannot describe high-level spatial location attributes of cells. It also cannot well reflect the openness of geographical space. Second, neighborhood definition based on geometric features cannot adapt to complex spatial interaction modes in real geographic systems, such as hierarchical diffusion and network diffusion. Unreasonable neighborhood definition will cause significant selection bias in simulation analysis. Third, the inherent assumption of cellular spatial homogeneity conflicts with the heterogeneity of actual geographical space. The traditional CA models cannot reflect the restrictive effects of economic distance and temporal distance on geographic flow. Fourth, most transition rules adopt globally deterministic and rely heavily on expert experience. Such rules cannot fully reflect the nonlinearity and randomness of geographic systems. Fifth, the high abstraction of the model ignores the combined effects of macroscopic environmental factors and multidimensional driving forces. This defect undermines the authenticity of simulation results. All the above limitations stem from the insufficient expression of geographic characteristics in classic CA model. Systematic expansion is required through the introduction of semantic information and multi-source data fusion. Territorial space belongs to a typical complex system. Its evolutionary process is jointly shaped by natural, social, economic, and policy factors. Territorial space exhibits prominent spatial heterogeneity and dynamic variations. The inherent flaws of the traditional CA models make it hard to identify the complex evolutionary features of territorial space, so their original frameworks cannot meet the practical simulation needs of TSUE. These practical demands drive the expansion and optimization of existing simulation models.

5.1.2 Machine Learning: Technical Challenges and Innovations for Simulation Optimization

The simulation limitations of the traditional CA models originate from the fundamental conflict between experience-driven simulation logic and the complex multifactor interaction features of land use evolution [3]. Machine learning serves as a data-driven technical approach [4, 5]. This method can independently mine complex internal relationships within research objects [6–8]. It provides a feasible solution to the above conflict and acts as a core technical support for the expansion and optimization of the CA models. The core purpose of introducing machine learning into the CA model simulation framework lies in making up for the inherent deficiencies of the traditional CA models with powerful data mining and predictive fitting capabilities. The integration process also faces multiple technical bottlenecks. The resolution of such challenges promotes the iterative upgrading of the overall simulation paradigm.

The primary technical difficulty lies in the effective integration between machine learning and the traditional CA models. Specific obstacles fall into two categories. The first obstacle relates to data compatibility. The traditional CA models mainly adopt spatial grid data as basic input. Most machine learning algorithms show higher adaptability to structured data [9]. The integration and standardized processing of spatial grid data and structured driving factor data form the foundation of model coupling. Reasonable processing modes can guarantee data consistency and overall validity. The second obstacle involves the fusion of evolutionary rules. Neighborhood rules of the traditional CA models follow clear spatial logic. Transition rules extracted by machine learning reflect data-driven nonlinear relationships. The organic combination of two types of rules becomes the core difficulty of model integration. Balanced design maintains the spatial evolution logic of the CA models. It also gives full play to the accurate fitting advantages of machine learning. Balanced design prevents separation between data-driven results and realistic spatial logic.

Existing studies have achieved key breakthroughs through targeted technical innovation. On the one hand, optimized data preprocessing technologies realize vectorization conversion and feature extraction of spatial grid data. Processed spatial data realizes correlation and matching with driving factor datasets. The integrated database meets the input requirements of mainstream machine learning algorithms and resolves fundamental data compatibility problems. On the other hand, current research establishes a dual rule system that combines machine learning derived rules and the traditional CA models spatial rules. Typical machine learning algorithms include random forest, support vector machine, and neural network. Such algorithms extract complex correlation between land type conversion and various driving factors from massive historical records. The output results generate objective and accurate cell state transition rules. Neighborhood rules of the original CA model remain intact throughout the operation process. These rules restrict the spatial interaction behavior of cellular units. The overall framework realizes organic unification between data-driven calculation and geographical spatial logic. Such technical progress reduces the subjectivity of artificial transition rules in the traditional CA models. It also improves the self-adaptation ability of simulation systems. The innovative mode promotes the overall transformation of the CA models from experience-driven operation to data-driven operation. Relevant research findings lay a solid technical foundation for the construction of the extended CA model.

5.1.3 Core Ideas of Extended CA Model: A Framework Combining Data Driving and Knowledge Guidance

In view of the inherent limitations of the traditional CA models and the technical progress brought by machine learning, the construction of the extended CA model breaks the single rule and experience-dominated logical framework of traditional simulation. Such models establish an integrated simulation system that combines data driving and knowledge guidance. The new framework achieves synchronous improvement in simulation accuracy, geographical logic, and practical guidance. It also forms the fundamental difference between the extended CA model and traditional ones.

The core philosophy of the extended CA model follows an integrated logic of dual support and dual constraint. Data driving acts as the primary support. Transition rules extracted by machine learning algorithms guarantee the accuracy and objectivity of simulation on TSUE. Research collects and organizes long-term historical records of TSUE in specific regions. Relevant datasets cover physical geography, socioeconomic conditions, and traffic information. Machine learning algorithms capture the internal evolutionary laws of TSUE through autonomous learning. Generated state transition rules fit regional characteristics well and replace traditional rules formulated by empirical judgment. Such improvement reduces human subjective interference effectively. It also enhances the model capacity to identify TSUE evolution under the interaction of complex multi-influencing factors. Knowledge guidance serves as the restrictive support throughout the simulation process. The integration of geographical theories and planning knowledge maintains geographical rationality and practical significance of simulation outputs. Geographical knowledge includes basic laws such as spatial non-stationarity and spatial dependence. The construction of spatial weight matrices and the adoption of spatial analysis methods restrict the spatial correlation of cell state transition. Relevant settings ensure simulation results of TSUE conform to objective geographical evolution laws. Planning knowledge involves territorial spatial planning policies and spatial regulation measures. Reasonable restrictive conditions avoid inappropriate land conversion within ecological protection red lines. Typical restrictions prohibit the transformation of inner cells into high consumption and low-efficiency urban construction land. Such designs align simulation outcomes with realistic planning demands and strengthen practical application value.

The integrated framework of the extended CA model consists of three core modules. The first module refers to the data layer. This layer completes the collection, cleaning, standardization, and integration of multi-source data. It focuses on the collation of indicators related to TSUE. A comprehensive database combining spatial data and attribute data provides solid basic support for model operation. The second module is defined as the rule layer. This layer integrates transition rules mined by machine learning and spatial neighborhood rules of classic CA models. Geographical and planning knowledge further formulates additional restrictive rules. A triple rule system covers data-based rules, spatial rules, and constraint rules. The systematic design adapts well to the evolutionary characteristics of TSUE. The third module represents the simulation layer. Iterative updating of cell states realizes dynamic simulation of TSUE under the established rule system. Embedded verification mechanisms adjust model parameters through comparison with historical records. Dynamic correction ensures stable and reliable simulation accuracy. This integrated framework retains the inherent spatial simulation advantages of the CA model. The combination of data driving and knowledge guidance makes up for multiple defects of the traditional CA models. It promotes the comprehensive upgrading of the existing simulation paradigm. The proposed theoretical and technical system provides a feasible solution for accurate simulation of TSUE in complex geographical scenarios.

5.2 Theoretical Basis: Overview of Core Model Principles

5.2.1 Spatial Feature Extraction: Fundamentals of CNN Model

The CNN model is a widely applied deep learning architecture, whose design originates from biological studies on the visual cortex of cat brains conducted by Hubel and Wiesel [10]. Subsequent research greatly promotes the improvement and optimization of the CNN model frameworks. Representative studies include LeNet-5 proposed by LeCun and relevant colleagues, and AlexNet developed by Krizhevsky and other scholars [11, 12]. These classical architectures drive the continuous development of the CNN model. As shown in → Fig. 5.1, the basic structure of the CNN model contains five key components. The main modules include input layer, the convolutional layer, pooling layer, fully connected layer and output layer. The input layer receives raw image data. The convolutional layer is responsible for extracting image features. Multiple convolution operations are conducted through a set of convolution kernels. Original input data can be converted into low dimensional feature vectors with strong representation capacity. The pooling layer undertakes the essential function of down sampling. Feature data output from convolutional layers contains substantial redundant information, which pooling layers eliminate via sampling and aggregation [13]. Key feature information can be well retained in the meantime. Such mechanisms effectively reduce data dimensionality [14]. Pooling layers are generally arranged between adjacent convolutional layers to connect different network modules. Fully connected layers are commonly placed in the final stage of network structures. These layers reorganize and fuse extracted features. The output layer generates differentiated results according to specific research objectives and application scenarios of convolutional neural networks [15]. Currently, the CNN model techniques have achieved mature application effects in multiple research fields [16]. Typical application directions cover image classification, image retrieval, object detection, and image segmentation. It also achieves competitive performance in natural language processing and other interdisciplinary research [17].

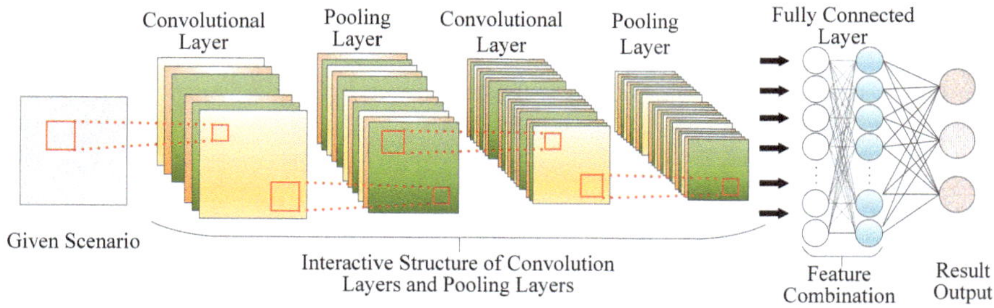

Fig. 5.1 Schematic diagram of convolutional neural network model structure

5.2.2 *Time-Series Dependency Modeling: Fundamentals of LSTM Model*

The LSTM model is an improved neural network model proposed by Hochreiter and Schmidhuber [18]. This model was developed to address the inherent limitations of the traditional recurrent neural networks (RNN) model [19]. Common defects include gradient explosion, gradient vanishing and insufficient long-term memory capacity during model operation. The LSTM model introduces memory cells along with forget gates, input gates, and output gates. Compared with the standard RNN model, such structural design strengthens the capacity to capture long-term dependencies and delayed information within time-series data. As illustrated in → Fig. 5.2, the LSTM model adopts a chain structure with four core network layers. The model framework involves two key memory mechanisms of long-term memory and short-term memory. Long-term memory is marked by red lines. Corresponding information passes through neural units without additional changes. Short-term memory is marked by black lines. Special gating structures support the update and output of short-term memory information [20]. This section elaborates on its working mechanism taking time step t as an example.

1. The forget gate controls information generated prior to time step t. This structural design filters out invalid and interfering information from early features. Joint inputs of h_{t-1} and X_t determine the retention or abandonment of information contained in C_{t-1}. The corresponding mathematical expression is presented as Eq. (5.1)

$$f_t = sigmoid\left(W_f \cdot \left[X_t, h_{t-1}\right] + b_f\right) \tag{5.1}$$

In the formula, W_f refers to the weight matrix of the forget gate, and b_f denotes the bias vector.

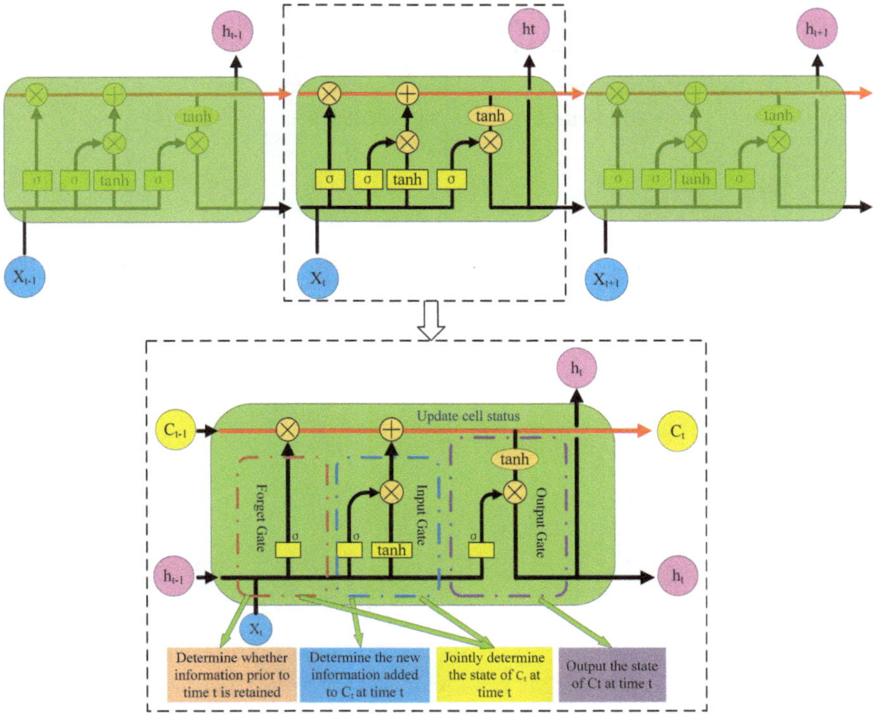

Fig. 5.2 Schematic diagram of long short-term memory model structure

2. The input gate determines new information added to C_t at time step *t*. This process takes h_{t-1} and X_t as joint inputs to the sigmoid function and the tanh function. The sigmoid function screens specific information for the update of i_t. The tanh function generates new candidate cell information C_t^*. The relevant mathematical expressions are presented as Eqs. (5.2, 5.3)

$$i_t = sigmoid\left(W_i \cdot \left[X_t, h_{t-1}\right] + b_i\right) \tag{5.2}$$

$$C_t^* = \tanh\left(W_c \cdot \left[X_t, h_{t-1}\right] + b_c\right) \tag{5.3}$$

In the formula, W_i and W_c denote the weight of the input gate and the weight for candidate cell calculation respectively. b_i and b_c represent corresponding bias vectors.

3. The forget gate and the input gate jointly determine the cell state C_t at time step *t*. The output f_t obtained from the forget gate conducts the Hadamard product

operation with C_{t-1} at the previous time step. The updated information i_t from the input gate performs the same operation with the candidate cell information C_t^*. The superposition of the two calculation results yields the cell state C_t at time step t. The mathematical expression is presented as Eq. (5.4)

$$C_t = f_t \odot C_{t-1} + i_t \odot C_t^* \tag{5.4}$$

4. The output gate completes the output of cell state C_t at time step t. In this process, the sigmoid function defines the gating state O_t of the output gate. The tanh function scales the cell state C_t for subsequent multiplication with O_t. Such calculation produces the final output of the model. The mathematical formulas are presented as Eqs. (5.5, 5.6)

$$O_t = sigmoid\left(W_o \cdot \left[X_t, h_{t-1}\right] + b_o\right) \tag{5.5}$$

$$h_t = O_t \odot \tanh(C_t) \tag{5.6}$$

In the formula, W_o indicates the weight of the output gate, and b_o refers to the bias vector.

Owing to the superior capacity of the LSTM model in processing long time-series information, this model has been widely applied in multiple research fields. Common applications include stock price forecasting, traffic flow prediction, air pollutant estimation, and natural disaster early warning [21–23].

5.2.3 Nonlinear Relationship Fitting: Fundamentals of RF Model

The RF model is a decision-tree-based machine learning algorithm proposed by Breiman [24, 25]. As shown in → Fig. 5.3, the model framework comprises three core procedures: sample set selection, decision tree construction, and voting integration [26–28]. First, sample set selection adopts the bootstrap resampling method. The method performs m sampling iterations with replacement from the original dataset D_m. A total of n sub-datasets can be constructed accordingly. Repetition appears among samples not only inside each sub-dataset but also between distinct sub-datasets. Second, decision trees are constructed using the resampled sub-datasets. For each tree node, b candidate features are randomly selected from the original A available features, where the value of b is smaller than A. The feature subspace formed by the selected b features guides the selection of classification features and split points for decision tree construction. Third, each individual decision tree acts as a base classifier for independent prediction. Final predictions are determined by majority voting across all tree outputs. Such integration mode determines the overall

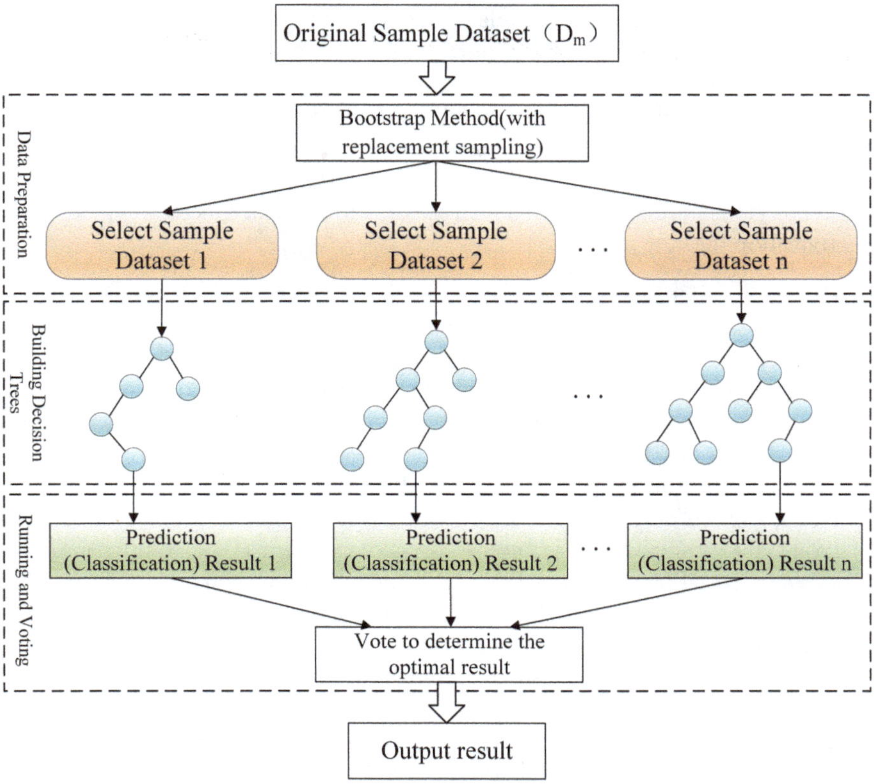

Fig. 5.3 Schematic diagram of random forest model structure

prediction results of the RF model. The RF model has achieved effective application in numerous research scenarios. Typical fields include text classification, remote sensing image classification, $PM_{2.5}$ concentration prediction, and biomass estimation. This algorithm exhibits distinct advantages such as strong error tolerance, low overfitting risk, and high prediction accuracy [27, 29, 30].

5.2.4 Dynamic Simulation Engine: Principles and Classical Structure of CA Model

The CA model was proposed by John von Neumann and Stanislaw M. Ulam [1, 2]. It is a grid dynamic model based on the spatial interaction of microscopic individuals and temporal causal relationships. This model features discrete time, space, and cell states. The mathematical expression is presented as Eq. (5.7)

$$A = (L_d, S, N, F) \tag{5.7}$$

In the formula, A denotes a cellular automaton system. L represents the cellular space, where the integer variable d refers to the dimension of the cellular space. S stands for the finite and discrete state set of cellular units. N indicates the combined states of a single cell and all cells within its neighborhood. F corresponds to the transformation rule.

As shown in → Fig. 5.4, a complete CA model system contains four key elements. These key elements cover cells and their states, cellular space, neighborhood range, and transition rules.

1. Cells and cell states. A cell, also defined as a unit or basic element, serves as the fundamental simulation unit of the CA model. Cell states reflect a series of attributes covering social, economic, and natural dimensions [31].
2. Cellular space. Cellular space refers to the spatial set of all distributed cells and the area for subsequent evolutionary simulation. This space can present geometric forms in arbitrary dimensions. Two-dimensional cellular space has diverse classification types [31, 32], as illustrated in → Fig. 5.5
3. Neighborhood range. The neighborhood range refers to the cell set distributed around the central cell defined by specific geometric rules. This component serves as the core influencing factor for the state evolution of the central cell. The neighborhood range of one-dimensional CA models is generally defined by radius constraints. Two-dimensional CA models mainly adopt two classic neighborhood types, namely the von Neumann type and the Moore type, as shown in → Fig. 5.6
4. Transition rules. Transition rules constitute the core of the CA model. Such dynamic functions determine the cell state at time $t + 1$ on the basis of the cell state and neighborhood conditions at time t. For the widely applied Moore neighborhood, the mathematical formula of transition rules is expressed as Eq. (5.8)

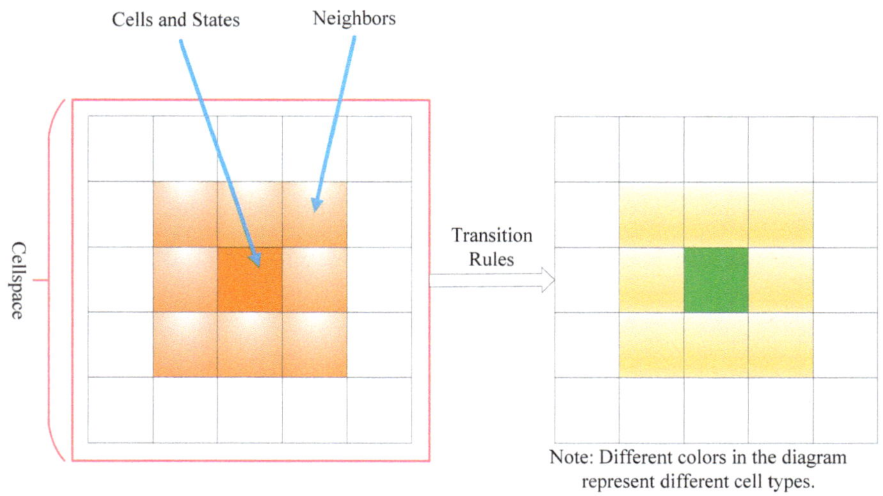

Fig. 5.4 Schematic diagram of cellular automata model structure

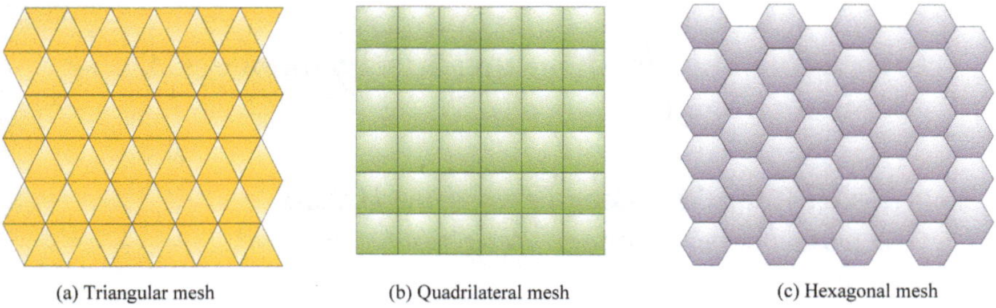

Fig. 5.5 Different types of spatial grids for two-dimensional cellular automata model

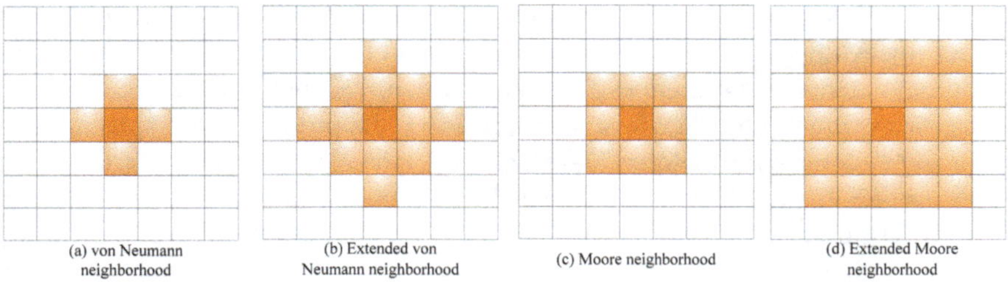

Fig. 5.6 Neighborhood Types of Cellular Automata Model

$$S_{ij}^{t+1} = f_N\left(S_{ij}^{t}\right) \tag{5.8}$$

In the formula, S denotes the cell state. f represents the transition function that defines the state change of cell ij from time t to time $t + 1$. N stands for the cell neighborhood and serves as the variable of the transition function.

The CA model adopts a bottom-up simulation paradigm. It possesses powerful spatial and parallel computing capabilities in spatial computation and parallel computing. These strengths bring distinct advantages for the simulation of complex spatiotemporal evolution characteristics [33]. Therefore, the CA model has gained extensive application in multiple research fields [34]. Typical research directions include topography and geomorphology, land use change, soil erosion, and natural disaster assessment [35].

5.3 Design of Extended Cellular Automata Model: Integrating Machine Learning into Traditional Cellular Automata Model

5.3.1 Model Coupling Idea: From Rule-Driven to Data-Driven Mode under Rule Constraints

The predictive modeling of TSUE is essentially a complex issue that involves dual dimensional modeling of time-series and spatial effects. From the perspective of time-series analysis, the evolution of TSUE is a gradual process with long-term temporal continuity. Efficiency change trends within long time-series exert decisive impacts on evolutionary trends. These trends reflect the historical context of TSUE evolution, determine its future evolutionary trajectories, and provide critical references for exploring efficiency change rules. From the perspective of spatial effect analysis, territorial space belongs to typical geographical spatial units. The evolution of its use efficiency is inevitably restricted by geographical laws and spatial effect theories. The efficiency change of a single territorial unit is not isolated but affected by spatial radiation from neighboring units. Such spatial interaction and correlation endow the evolution of TSUE with obvious complexity and dynamic characteristics. For the above reasons, the prediction of TSUE needs to comprehensively consider the combined influences of temporal factors and spatial factors.

In traditional CA models, the formulation of appropriate transition rules lies at the core of overall model operation. The state of neighboring cells acts as a decisive factor in judging cell conversion at the next time step during the formulation of such transition rules. Existing studies on the CA model transition rules reveal notable limitations from the time-series dimension. Neighborhood cell states only reflect spatial conditions at a single time section. Such static spatial representation fails to capture continuous cell transitions over long time series. From the spatial effect dimension, traditional CA models adopt statistical methods to calculate the proportional distribution of neighboring cell states. Such simple calculation modes help measure the transition probability of each cell. These methods neglect the complex spatial heterogeneity among geographical units. Inadequate characterization of spatial heterogeneity and constraints restricts the overall accuracy of model prediction results. The above analysis proves the inherent defects of traditional CA models in transition rule formulation. These models cannot fully accommodate the dual effects of temporal and spatial factors required by the predictive modeling of TSUE. This study proposes an innovative improvement scheme to address such deficiencies. This framework employs the CNN model for spatial feature extraction and the LSTM model for long time-series temporal feature capture; the derived spatiotemporal features are further input into the RF model as core explanatory variables. The RF model completes the quantitative calculation of cell transition probability. The optimized CA model finally realizes the dynamic prediction of future TSUE. The proposed coupling framework achieves more effective capture of temporal and spatial impacts on TSUE. This integrated method further enhances the overall accuracy and stability of relevant prediction models.

5.3.2 Model Coupling Mechanism: Multi-level Modular Fusion Architecture

The coupling mechanism of the extended CA model is presented in → Fig. 5.7. This coupled model takes the CA model as the core framework. Machine learning models including the CNN, LSTM, and RF models are introduced as auxiliary components. This integrated structure not only improves the prediction accuracy of TSUE but also helps identify its evolutionary trends. The extended CA model consists of three core functional modules. First, the CNN-LSTM model captures long time-series temporal and spatial features of neighborhood cells. This module focuses on the extraction of multi-dimensional spatiotemporal characteristics within continuous time sequences. Second, the RF module recognizes the transition suitability of TSUE. This module integrates multiple influencing factors of efficiency evolution. These driving factors include the spatiotemporal features derived from the preceding module. Third, the CA model performs final TSUE prediction. This module combines output results from the first two modules to complete high-precision simulation and prediction. This study adopts a typical sequential data coupling mode for model connection. The operation of a single model depends on the outputs of the preceding sub-models. As displayed in → Fig. 5.7, the CA model operates based on transition suitability data from the RF model. The RF model further requires spatiotemporal feature data captured by the CNN-LSTM model. Such a coupling structure ensures effective data flow among all sub-models. It maintains good coherence and overall accuracy throughout the whole prediction process of TSUE.

To verify the effectiveness of the extended CA model and conduct in-depth exploration, this study carries out accuracy validation with historical data of TSUE. The research collects relevant datasets for the years 1990, 2000, 2010, and 2020 from previous research results. The calculation process follows the indicator screening method in → Chap. 3 and the GeoDEA model measurement model in → Chap. 4. Core data logic remains consistent with the evaluation system of PLES efficiency in → Chap. 6. → Chap. 6 presents complete indicator systems and comprehensive efficiency values. Current datasets are less comprehensive and refined due to different research stages. This study analyzes the flow and conversion process of data across multiple models. The analysis clarifies parameter transmission paths and coupling logic between different modules. Relevant conclusions promote systematic understanding of the internal operation mechanism of the extended CA model. They also provide theoretical foundations and practical references for subsequent structural

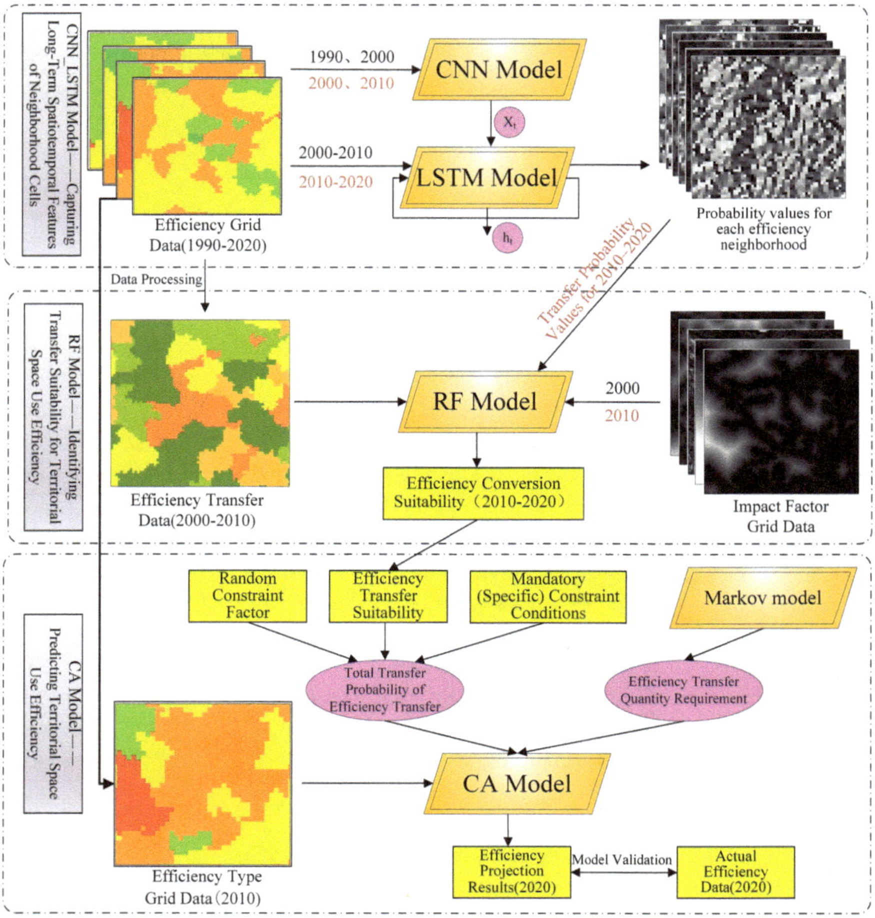

Fig. 5.7 Schematic diagram of the coupling mechanism for extended cellular automata model

optimization and prediction improvement. The extended CA model presents prominent universality and robustness. The model can maintain robust operational stability and reliable prediction performance after the replacement of research objects. Available replacement materials include similar datasets from other regions and time periods, as well as the complete efficiency data adopted in → Chap. 6. Such superior characteristics endow the extended CA model with broad application potential and research value.

The following part elaborates on three core modules of the extended CA model in detail.

1. **Long time-series spatiotemporal feature capture module of neighborhood cells based on the CNN-LSTM model**

This module focuses on the extraction of long time-series spatiotemporal features for cell units. The specific implementation process is illustrated in → Fig. 5.8. The CNN model adopts a 3 × 3 convolution window to slide over grid data of TSUE in 1990 and 2000. This operation extracts spatial features of grid cells and constructs a standard dataset of neighborhood spatial features. Grid calculation and spatial assignment are conducted on TSUE data of 2000 and 2010. These operations generate efficiency transition data covering the period from 2000 to 2010.Spatial feature datasets and efficiency transition datasets are converted between grid format and array format. All processed data undergo segmentation and grouping. Processed neighborhood spatial features serve as feature datasets. Corresponding efficiency transition results are taken as label datasets. Two types of data are imported into the LSTM model for parameter calibration and training. The modeling process further completes the extraction of temporal features. The fully connected layer outputs integrated spatiotemporal features acquired through joint model learning. Calculated neighborhood probability values are recorded accordingly. The trained model is stored steadily to support subsequent data calculation and analysis. The overall modeling process can be extended for multi-period simulation. Neighborhood spatial feature data of 2000 and 2010 support the prediction of neighborhood probability values for efficiency transition during 2010 to 2020. Neighborhood spatial feature data of 2010 and 2020 can generate predicted neighborhood probability values of efficiency transition for the period from 2020 to 2030.

2. **Transition Suitability Identification Module of TSUE Based on RF Model**

This module aims to calculate the transition probability of cell type conversion. One specific cell type can shift to other categories under comprehensive constraint conditions. The detailed workflow is presented in → Fig. 5.9. Geographic influencing factors and economic influencing factors in 2000 are integrated with output results of the CNN-LSTM model for the period 2000 to 2010. All the above indicators constitute independent variables for model training. Transition data of TSUE from 2000 to 2010 are calculated and set as dependent variables. Independent variables and dependent variables are combined and grouped before input into the RF model. The model constructs decision trees to complete parameter calibration and systematic training. Voting mechanisms inside the RF model generate final transition suitability data of efficiency changes from 2000 to 2010. The well-trained model is stored for subsequent data processing and simulation. The analytical framework can be applied to other research periods in the same logical manner. Geographic and economic factors, together with the 2010 outputs of the CNN-LSTM model, are used to derive transition suitability data for the period 2010–2020. Similarly, factors from 2020 and the corresponding CNN-LSTM model outputs generate predicted TSUE transition suitability data for 2020–2030.

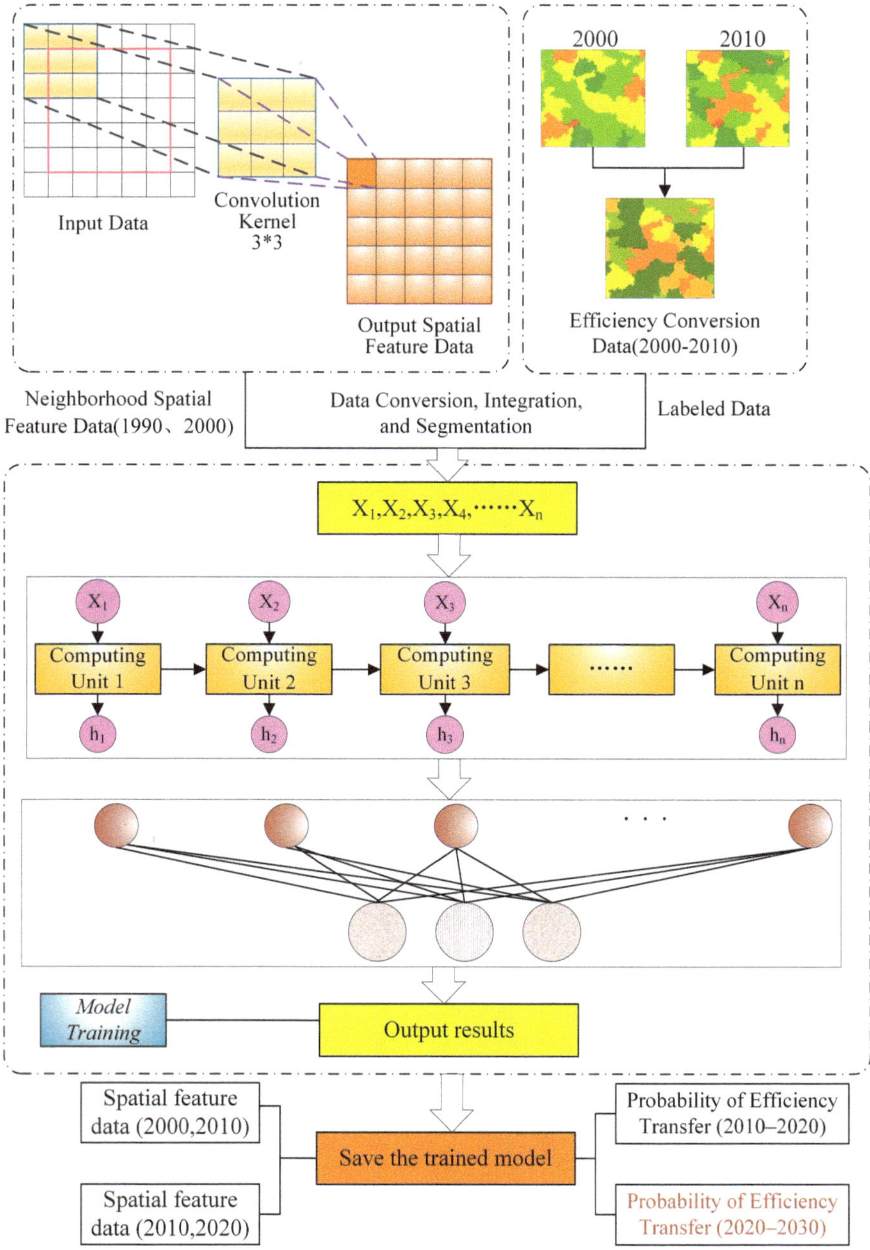

Fig. 5.8 Flow chart of convolutional neural network-long short-term memory model operation

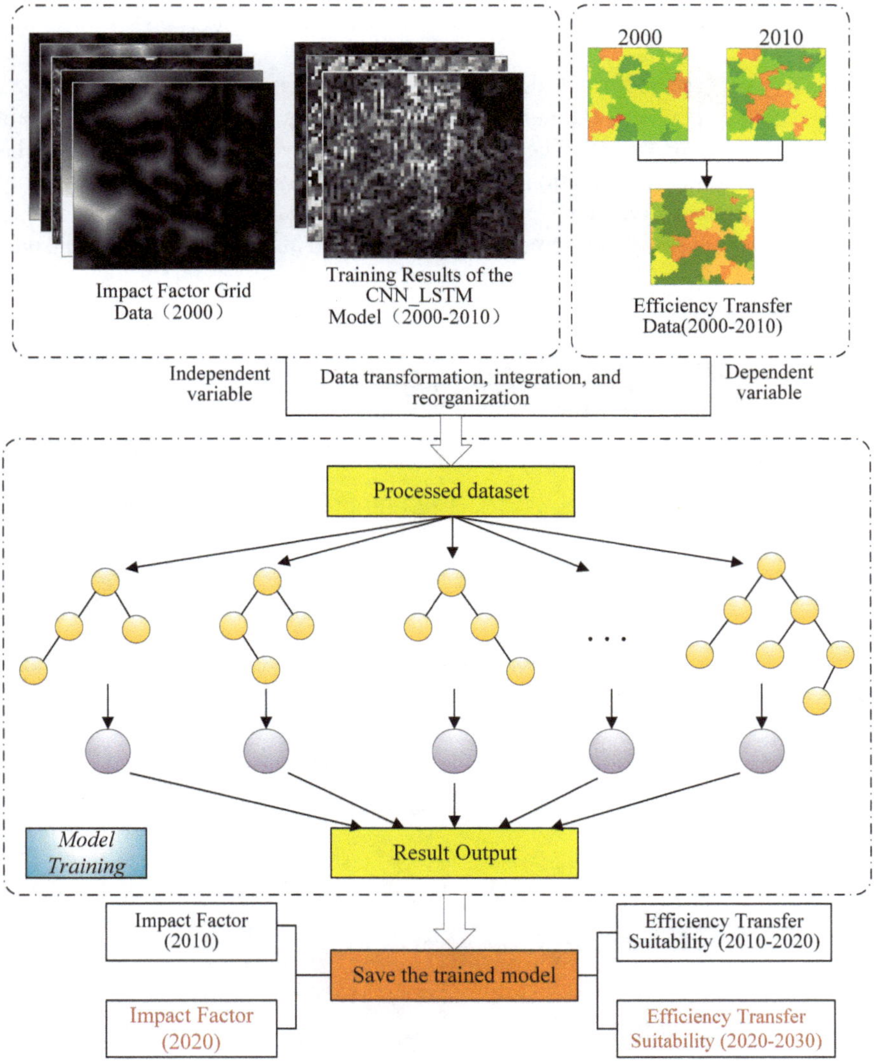

Fig. 5.9 Flow chart of random forest model operation

3. **Prediction Module of TSUE Based on CA Model**

This module undertakes the prediction of TSUE. The specific implementation process is shown in → Fig. 5.10. Three types of basic data support the stable operation of this module. The first type refers to grid data of TSUE in the previous phase of the target prediction year. The second type represents the transition probability of conversion between different cell types. Such probability values originate from the joint output of the CNN-LSTM model and the RF model. Core components include transition suitability data, random influencing factors, and mandatory constraint factors.

The third type refers to the number of cells to be converted, calculated via the Markov model.

All basic data are integrated and imported into the CA model. The model conducts repeated cyclic operations through random cell selection. Major procedures contain quantity comparison of efficiency grids, category conversion, quantity recalculation, and probability recalculation. Circulation operations continue until the quantity demand of efficiency type conversion reaches the standard threshold. Final prediction results are exported and stored in complete form. This study adopts original data

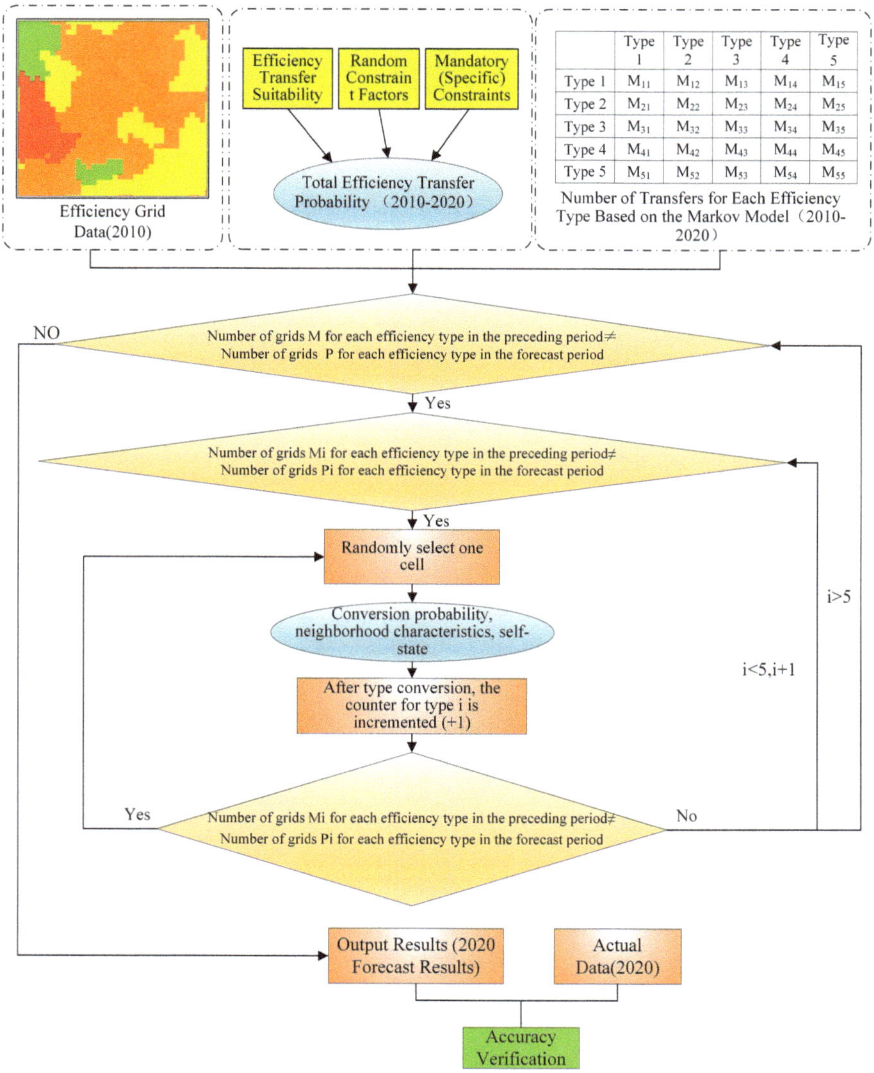

Fig. 5.10 Flow chart of cellular automata model operation

in 2010 and efficiency transition suitability data from 2010 to 2020 generated by the RF model. These materials support the simulation and prediction of efficiency types in 2020 as well as subsequent model verification.

5.4 Implementation and Validation of Extended Cellular Automata Model

5.4.1 Design and Implementation of Model Validation Experiments

This study formulates a targeted experimental framework to verify the effectiveness of the extended CA model. With high-performance computer hardware and professional software development environments, self-compiled codes implement the full workflow of batch data processing, model construction, and model operation. The experiment is designed based on rigorous logic. Training datasets include TSUE data and fifteen categories of influencing factors for 340 Chinese cities across 1990, 2000, and 2010, which are fed into the extended CA model for systematic training. The model learns the evolution rules, influencing mechanisms and spatiotemporal characteristics of TSUE from 1990 to 2010. It further generates predicted data of national TSUE for 2020. Multi-dimensional comparative analysis is conducted between predicted efficiency data and actual measured data of TSUE in 2020. Quantitative error calculation and consistency assessment serve as the core evaluation indicators. Comparative results finally support the effectiveness verification of the extended CA model. The verified model can accurately capture the evolution characteristics of TSUE. Reliable technical and methodological support can be provided for subsequent simulation and prediction of future efficiency changes.

Regarding the professional software development environment, this study adopts the PyCharm 2023.2.1 Community Edition integrated development platform. Python 3.10 is applied for data processing, model construction, and model operation. In the batch data processing stage, multiple toolkits underpin data preprocessing. Major toolkits include the numpy 1.25.2 scientific computing library, the osgeo 3.6.2 open-source geodatabase library, and the pyrsgis 0.4.1 geospatial data processing library. These tools support batch preprocessing of socioeconomic data and geospatial data in 1990, 2000, and 2010. Specific operations include data import, type conversion, integration, and standardization. The preprocessing procedure ensures the accuracy, consistency, and integrity of training data. A solid foundation is thus established for subsequent model training.

The training and operation of the extended CA model include three core collaborative modules. The coordinated operation of all modules guarantees the smooth progress of the experimental process. The first module corresponds to the implementation of the CNN-LSTM model. It first defines relevant data processing functions to integrate, classify and segment training and label datasets from 1990 to 2010, thereby extracting long time-series spatiotemporal features of neighborhood cells. The CNN-LSTM model class under the paddle 2.5.1 deep learning framework is then invoked. Related parameters are adjusted to meet experimental requirements. The module achieves accurate capture of spatiotemporal features. The second module represents the implementation of the RF model. This module is built on the sklearn.ensemble sub-module in the sklearn 1.2.2 machine learning library. The RandomForestClassifier class is imported for model construction. Spatiotemporal features extracted by the CNN-LSTM model and 15 types of influencing factors serve as model inputs. The trained module identifies the transition suitability of TSUE. The third module represents the implementation of the CA model. This module relies on multiple Python toolkits including numpy 1.25.2, the standard time library, and the random number library. Output results from the first two modules are integrated into this module. A series of cyclic operations are conducted, including cell selection, type conversion, and quantity calculation. The module completes the prediction of TSUE data in 2020. Final prediction results are exported for comparison and verification with actual measured data.

5.4.2 Data Preprocessing and Parameter Configuration

5.4.2.1 Data Preparation and Preprocessing

1. **Preparation and Preprocessing of TSUE Data**

The TSUE data adopted in this chapter are calculated by the models constructed in → Chap. 3 and → Chap. 4. Complete data results are presented in → Chap. 6. Only partial data are displayed in this section to verify the feasibility of the extended CA model. The efficiency values measured by the GeoDEA model range from 0 to 3.6. Efficiency values vary noticeably across territorial units, which creates challenges for in-depth analysis, comparative research, and quantitative prediction. This study analyzes the distribution characteristics of efficiency values and divides all samples into five hierarchical development stages, as listed in → Table 5.1. With the support of ArcGIS Pro 3.0, segmented classification is conducted on PLES use efficiency and overall TSUE of 340 prefecture-level cities across China in 1990, 2000, 2010, and 2020. Five category codes, ranging from 1 to 5, are assigned according to corresponding value intervals. Vector data are finally converted into raster data for subsequent spatial modeling. This study conducts detailed analysis and experimental demonstration to determine the appropriate raster resolution. The research scope covers 340 cities across 29 provincial-level administrative regions nationwide. The study area covers a wide geographic scope. A 1 km × 1 km raster resolution would generate massive grid data, including 5185 columns, 3702 rows, and a total of 19,194,870 grids. Excessive data volume would substantially increase computational burden. Model operation would rely entirely on high-end computing equipment. Moreover, the basic evaluation unit of this study refers to municipal administrative units rather than single raster cells. Each city matches one unified efficiency value in original measurement results. The future efficiency evolution of one city is strongly affected by the efficiency status of surrounding cities. Under the raster scale, grid units within urban interiors show weak variation potential. Grids distributed at urban boundaries present obvious conversion tendencies. Irregular boundaries between adjacent cities can be fully identified under both 1 km and 10 km raster resolutions. A 1 km resolution leads to excessive refinement of internal urban grids. Limited variation characteristics of interior grids bring limited practical significance for model simulation. Combining the above two considerations, all spatial data are resampled to a unified 10 km by 10 km resolution. This resolution balances computational efficiency and spatial expression accuracy for model training and operation. Predicted efficiency results are generated on the raster basis. The average value of all raster grids within each municipal boundary is calculated as the final predicted efficiency of the corresponding city. Subsequent verification proves that the model accuracy meets the overall research requirements.

2. **Preparation and Preprocessing of Influencing Factor Data**

As shown in → Table 5.2, the prediction of TSUE involves a total of 15 influencing factors. Python programming and ArcGIS Pro 3.0 software are adopted to complete the preparation and preprocessing of influencing factor data. This process ensures the accuracy and consistency of all datasets. First, raster projection processing is conducted on all acquired data. The projected coordinate system of each dataset is uniformly adjusted to WGS_1984_UTM_Zone_50N. The geographic coordinate system is uniformly set to WGS-84. These operations ensure subsequent spatial analysis. Next, five key influencing factors are processed via clipping and reclassification. These factors include DEM, gross production distribution, population distribution, annual precipitation, and annual average temperature. Slope and aspect datasets are calculated independently based on original DEM data. In addition, Euclidean distance tools in spatial analysis are applied to generate eight distance factors. These factors include the distance to highways, railways, water bodies, primary schools, middle schools, universities, hospitals, and urban centers. A series of preliminary preparations are required before spatial calculation. Typical operations include the extraction of railway and highway layers from vector files according to attribute fields. Data clipping is conducted based on the boundary of the study area. Notably, partial data

Table 5.1 Definition and classification coding of territorial space use efficiency grades

Efficiency value interval	[0,0.36)	[0.36,0.72)	[0.72,1.08)	[1.08,1.44)	[1.44,3]
Category code	1	2	3	4	5
Efficiency level	Low-efficiency Level	Medium-low Efficiency Level	Medium-efficiency Level	Medium-high Efficiency Level	High-efficiency Level

are missing due to the long research span. Data from adjacent years are used to replace missing records in the research process. For instance, gross production data in 1992 and 2019 replace the records in 1990 and 2020, respectively. This substitution method ensures data continuity and integrity. No significant impact is exerted on the final prediction results. Furthermore, several factors remain stable during the whole research period. DEM, slope, aspect, distance to water bodies, and distance to urban centers share a unified dataset. The same processing strategy is also applied to the datasets for distances to primary schools, middle schools, universities, and hospitals. This operation follows existing research frameworks and accommodates data availability constraints. Finally, the resolution of all influencing factor rasters is unified to 10 km × 10 km.

5.4.2.2 Key Parameters of Extended CA Model

1. **Main Parameter Setting of the CNN-LSTM Model**

As shown in → Fig. 5.11, the CNN-LSTM model constructed in this study contains two convolutional layers, two pooling layers, one dropout layer, one LSTM layer, and one fully connected layer. The convolution kernel is set to 3 × 3, and the sliding stride is defined as 1.

Convolution kernel size directly affects the extraction performance of spatial feature information. Small kernels present limited capability in feature extraction and bring higher memory consumption. Large kernels benefit the capture of global and macroscopic features but show poor performance in identifying shallow local features. Research units in this study differ greatly in area. The smallest territorial unit covers approximately 500 square kilometers. The unified raster resolution is 10 km × 10 km. The smallest territorial unit thus contains nearly five grid cells. Repeated trials and comprehensive comparison support the selection of a 3 × 3 convolution kernel. This configuration captures detailed change information of TSUE. It also avoids operational stagnation caused by excessive data volume. The sliding stride is set to 1 to fully retain spatial feature information of each grid cell. The sliding process starts from the upper left corner of input data and proceeds in a single-step mode. Pooling layers adopt a 2 × 2 window to reduce data dimensionality and compress overall data volume. The dropout layer is added to mitigate model overfitting. According to output results of the CNN model and experimental verification, the number of LSTM layers is set to 1, and the hidden layer size is defined as 16. Several training parameters are determined for model optimization. The batch size is set to 256, and the training epoch is fixed at 100.

2. **Main Parameter Setting of RF Model**

Repeated experiments, verification, and comprehensive trade-offs are combined with existing research findings in parameter calibration. The random seed is set to 10 to ensure experimental reproducibility. The number of decision trees within the random forest is defined as 100. The parameter for parallel processing control is set to

Table 5.2 Detailed information list of influencing factor data

No.	Influencing Factor	Original Data Source	Original Data Format	Year	Data Preprocessing Method
1	DEM	Resource and Environment Science and Data Center (https://www.resdc.cn/)	Raster Data	2000	Coordinate system conversion, clipping, resampling, etc.
2	Slope	–	–	–	Calculated based on DEM data
3	Aspect	–	–	–	Calculated based on DEM data
4	GDP Distribution	Resource and Environment Science and Data Center (https://www.resdc.cn/)	Raster Data	1992 (instead of 1990), 2000, 2010, 2019 (instead of 2020)	Coordinate system conversion, clipping, resampling, etc.
5	Population Distribution	Resource and Environment Science and Data Center (https://www.resdc.cn/)	Raster Data	1990, 2000, 2010, 2019 (instead of 2020)	Coordinate system conversion, clipping, resampling, etc.
6	Annual Precipitation	Resource and Environment Science and Data Center (https://www.resdc.cn/)	Raster Data	1990, 2000, 2010, 2020	Coordinate system conversion, clipping, resampling, etc.
7	Annual Average Temperature	Resource and Environment Science and Data Center (https://www.resdc.cn/)	Raster Data	1990, 2000, 2010, 2020	Coordinate system conversion, clipping, resampling, etc.
8	Distance to Highways	Obtained from the official OpenStreetMap website	Vector Data	2020	Coordinate system conversion, attribute extraction, clipping, Euclidean distance calculation
9	Distance to Railways	Obtained from the official OpenStreetMap website	Vector Data	2020	Coordinate system conversion, attribute extraction, clipping, Euclidean distance calculation
10	Distance to Water Bodies	Obtained from the official OpenStreetMap website	Vector Data	2019	Coordinate system conversion, Euclidean distance calculation, etc.

(continued)

Table 5.2 (continued)

No.	Influencing Factor	Original Data Source	Original Data Format	Year	Data Preprocessing Method
11	Distance to Primary Schools	Amap	POI Point Data	2019	Coordinate system conversion, Euclidean distance calculation, etc.
12	Distance to Middle Schools	Amap	POI Point Data	2019	Coordinate system conversion, Euclidean distance calculation, etc.
13	Distance to Universities	Amap	POI Point Data	2019	Coordinate system conversion, Euclidean distance calculation, etc.
14	Distance to Hospitals	Amap	POI Point Data	2019	Coordinate system conversion, Euclidean distance calculation, etc.
15	Distance to Urban Centers	Amap	POI Point Data	2019	Coordinate system conversion, Euclidean distance calculation, etc.

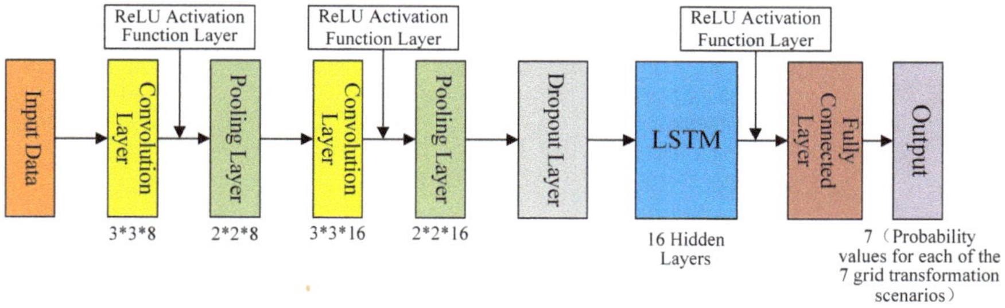

Fig. 5.11 Structural diagram of the convolutional neural network-long short-term memory model

10 to support simultaneous processing of ten tasks. A total of 20 features are selected as model inputs for RF construction. The input variables cover 15 influencing factors and five conversion probabilities corresponding to efficiency level types.

3. **Main Parameter Setting of CA Model**

The Moore neighborhood with a radius of 1 is adopted in the operation of the CA model. The central cell receives state influences from eight surrounding cells in vertical, horizontal, and diagonal directions. Other core parameters are presented in detail in → Fig. 5.10 of → Sect. 5.3.2. Relevant contents will not be repeated in this section.

5.4.3 Model Validation: Multi-Dimensional Evaluation of the Coupled Model

5.4.3.1 Evaluation Indicator System

This study adopts three indicators to verify the accuracy of the extended CA model, including the Kappa coefficient, F1-score, and Figure of Merit (FOM) coefficient. The Kappa coefficient, a confusion-matrix-based metric ranging from 0 to 1, is widely used for accuracy evaluation. A value above 0.6 is generally recognized as an indicator of reliable prediction. Predicted data maintain high consistency with actual observations, and relevant simulation results can support further scientific analysis [36]. The Kappa calculation formula is specified as Eqs. (5.9, 5.10, 5.11).

$$Kappa = \frac{p_o - p_e}{1 - p_e} \tag{5.9}$$

$$p_o = \frac{\sum_{i=1}^{5} w_i}{n} \tag{5.10}$$

$$p_e = \frac{\sum_{i=1}^{5} a_i \times b_i}{n \times n} \tag{5.11}$$

where n represents the total number of samples, and i denotes the grade type of TSUE. w_i refers to the number of samples with consistent actual and predicted efficiency grades of type i. a_i indicates the actual sample count of efficiency grade i, and b_i represents the predicted sample count of efficiency grade i.

F1-score serves as a common statistical indicator for the quantitative evaluation of model accuracy. Four core concepts are defined as follows. TP (True Positive) refers to correct positive discrimination results of the model. FP (False Positive) refers to incorrect positive discrimination results. TN (True Negative) refers to correct negative discrimination results. FN (False Negative) refers to incorrect negative discrimination results. F1-score is derived from four metrics: Accuracy, Precision and Recall. The value of all indicators ranges from 0 to 1. Higher values correspond to stronger prediction accuracy of the model [4]. Relevant calculation formulas are presented as Eqs. (5.12 and 5.15).

$$Accuracy = \frac{TP + TN}{TP + TN + FP + FN} \tag{5.12}$$

$$Precision = \frac{TP}{TP + FP} \tag{5.13}$$

$$Recall = \frac{TP}{TP + FN} \tag{5.14}$$

$$F1 - score = \frac{2 \times Precision \times Recall}{Precision + Recall} = \frac{2 \times TP}{2 \times TP + FP + FN} \tag{5.15}$$

The FOM (Figure of Merit) coefficient evaluates prediction performance by measuring discrepancies between actual observations and simulated results. Ranging from 0 to 1, the FOM coefficient positively correlates with model prediction performance [37]. The relevant calculation formula is presented as Eq. (5.16).

$$FOM = \frac{B}{A + B + C + D} \tag{5.16}$$

where A refers to the area of incorrect transition zones with actual land conversion and unpredicted changes. B represents the area of correct transition zones with consistent conversion in actual situations and prediction results. C indicates the area of unreasonable transition zones with predicted conversion and no actual change. D stands for the area of stable zones without conversion in both actual conditions and model prediction.

5.4.3.2 Comparative Analysis of Model Accuracy

This study adopts the above three core indicators to perform comparative analysis between the extended CA model and the traditional CA models. The comparison aims to further verify the prediction accuracy and stability of the extended CA model. Both models are applied to simulate TSUE data in 2020. All predicted outcomes are compared and analyzed against measured data of the same year. Detailed comparison results are presented in → Table 5.3. The Kappa coefficient of the extended CA model reaches 0.8116, with an increase of approximately 6.9 percentage points compared with the traditional CA model. Such outcome fully proves the superior prediction performance of the extended CA model. Four additional evaluation indicators include accuracy, precision, recall, and F1-score. The traditional CA model yields a marginally higher Recall value by 0.33 percentage points. The extended CA model still achieves higher scores in accuracy, precision, and F1-score. Meanwhile, the FOM value of the extended CA model is 11.22 percentage points higher than that of the traditional model. → Fig. 5.12 shows the visual comparison of prediction results from the two models. The simulation output of the extended CA model maintains significantly better consistency with actual measured data. In general, the extended CA model exhibits distinct advantages across multiple accuracy indicators and possesses much stronger prediction capability. The extended CA model is more applicable to the future prediction of TSUE, and can generate more accurate and reliable simulation outputs for relevant research.

Table 5.3 Accuracy comparison between extended cellular automata model and traditional cellular automata model

Model	Kappa	Accuracy	Precision	Recall	F1-score	FOM
Traditional CA model	0.7426	0.8159	0.8006	0.8089	0.8021	0.3984
Extended CA model	0.8116	0.8541	0.8224	0.8056	0.8275	0.5106

5.4.4 Comprehensive Conclusions

Based on multi-dimensional accuracy verification and comparative tests, the core findings of the extended CA model are summarized as follows. First, the data preprocessing workflow of the extended CA model is scientific and reliable. The grid resolution is set at 10 km × 10 km. This configuration balances processing efficiency and prediction accuracy, while aligning with the municipal-scale measurement logic of the GeoDEA model in → Chap. 4. The workflow can fully meet the research demands for the prediction of TSUE. Second, key modules of the extended CA model include the CNN-LSTM, RF, and CA models. Parameter configurations of all modules are verified through repeated trials. Such configurations exhibit good repeatability and rationality, and the entire model maintains stable operation performance. Third, results from multi-dimensional accuracy assessment confirm outstanding performance of the extended CA model. Evaluation indicators cover the Kappa coefficient, F1-score, and FOM coefficient. The proposed model achieves higher prediction accuracy than the traditional CA model and presents high consistency with actual measured data. Fourth, the extended CA model integrates machine learning algorithms and the traditional CA framework. This integration overcomes inherent defects of traditional CA models. The improved model can effectively capture the temporal and spatial evolution characteristics of TSUE and present prominent practical values.

5.5 Chapter Summary

5.5.1 Characteristics and Contributions of the Extended CA Model

This chapter focuses on the trend prediction of TSUE. Relevant research completes the construction, implementation, and verification of the extended CA model. This model breaks the inherent limitations of traditional CA models that rely on empirical driving and single rules. It establishes an innovative simulation paradigm supported by data driving and knowledge guidance. The proposed model presents distinct innovation and practicality. In terms of technical features, the extended CA model takes the traditional CA model as the core of dynamic simulation. It couples three machine

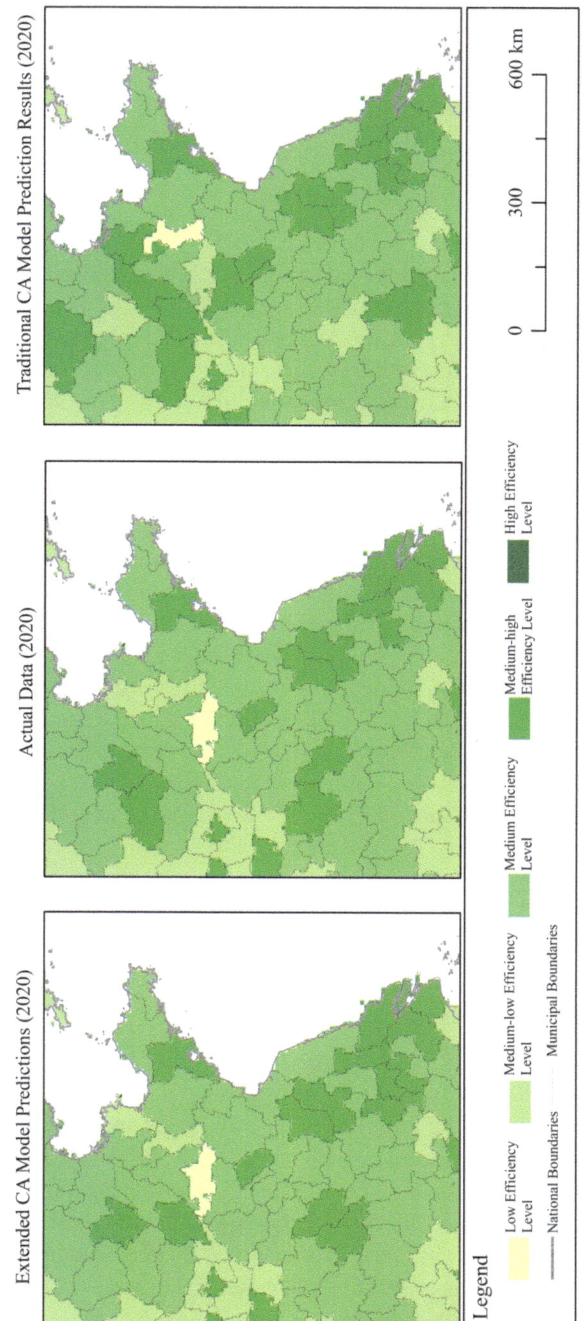

Fig. 5.12 Comparison of prediction results: extended cellular automata model and traditional cellular automata model (Partial display)

learning models: CNN, LSTM, and RF. This framework realizes the accurate extraction of temporal and spatial features. It also supports the effective fitting of nonlinear relationships and the scientific calculation of transition probability. The new model retains the bottom-up advantages of the traditional CA models in spatial simulation. It also adopts machine learning methods to compensate for the weaknesses of traditional CA models in geographic feature expression and multi-factor interaction identification. From the perspective of data and rules, this study constructs a comprehensive dataset containing spatial data and attribute data. It establishes a three-level rule system composed of data rules, spatial rules, and constraint rules. The framework integrates geographic knowledge and planning and policy constraints. Such optimization improves simulation accuracy, geographic rationality, and practical guiding significance in a coordinated manner.

The core contributions of this chapter lie in three dimensions. First is the theoretical progress. The research promotes the paradigm shift of the CA model from empirical driving to data driving and rule constraint. This study improves the theoretical framework for predicting TSUE under complex geographic scenarios. The research also enriches existing findings on the coupled application of machine learning and the traditional CA models. Second is the technical innovation. This research constructs a multi-level and modular coupling framework of CNN-LSTM-RF-CA model. The new framework solves key defects of traditional CA models. Common defects include insufficient geographic feature expression, strong subjectivity in conversion rules, and weak adaptive capacity. This study proposes a replicable and generalized technical route for the prediction of TSUE. Third is the practical value. Empirical tests cover 340 cities across the country. The test results prove the high accuracy and stability of the extended CA model. The model delivers far better prediction performance than traditional CA models. Relevant outcomes provide solid technical support for dynamic monitoring, trend prediction, and planning optimization of TSUE.

5.5.2 Application Scenarios and Notes of the Extended CA Model

Combined with model construction logic and empirical verification results, the extended CA model shows clear targeting in applicable scenarios. This model serves dynamic simulation and trend prediction research of TSUE at large scales and with long time sequences. The model fits regions with vast territory and obvious spatial heterogeneity. Such regions are comprehensively affected by natural, social, economic, and policy factors. The extended CA model can be widely applied in multiple fields. Typical fields cover territorial spatial planning, land use optimization, and regional development decision making. Relevant results deliver accurate reference of efficiency evolution trends for policymakers. The extended CA model also presents favorable universality. Simple data replacement and parameter fine adjustment can enable the model to adapt to prediction demands of TSUE in regions and scales of different types. The model can also expand to simulation research on the evolution of other complex geographic systems with similar attributes.

Attention shall be paid to several key points in the application of the extended CA model. First lies in the standardization of data preprocessing. Model operation sets strict requirements for data integrity and consistency. Unified projection and coordinate systems need standardized implementation across all datasets. Reasonable solutions should be adopted to address missing data. Standardized grid resampling and feature engineering help maintain qualified data quality for model training and simulation. Second refers to the rationality of parameter configuration. Hyperparameters of CNN-LSTM model, RF model and CA sub-models shall match regional features and data characteristics. Repeated experimental verification and comprehensive evaluation support scientific parameter selection. Proper configuration prevents the decline of simulation accuracy caused by unreasonable parameter combination. Third concerns the adaptability of constraint rules. The introduction of geographic knowledge and planning policies must conform to local realistic conditions. Typical constraints include ecological protection red lines and management regulations of territorial spatial planning. Practical alignment endows simulation results with reliable guiding value for actual management. Fourth reflects the rational matching of research scales. The setting of grid size balances research scope and overall data volume. Excessively refined or simplified grids will cause systematic deviation in simulation outputs. Reasonable grid specifications maintain spatial matching between basic analysis units and research objects such as municipal administrative regions.

5.5.3 Limitations and Future Research Directions of the Extended CA Model

The extended CA model achieves high accuracy and practical performance in the prediction of TSUE. The model still has certain limitations based on research procedures and empirical results. Such limitations require further discussion and targeted optimization. First, the overall coupling depth of the model needs further improvement. The current framework uses sequential data coupling, leading to insufficient interaction between sub-modules. Temporal and spatial features extracted by the CNN-LSTM model module lack sufficient coordination with transition probability calculation of the RF model and cellular iteration of the CA model. The whole system fails to realize in-depth integration of respective technical advantages. Second, the selection of driving factors shows obvious restrictions. The research incorporates 15 influencing factors covering natural, social, and economic dimensions. The model ignores the disturbance from atypical elements. Typical examples include extreme climate events, sudden policy adjustments, and major infrastructure construction.

Such omitted factors exert obvious impacts on TSUE. The model presents insufficient response capacity for short-term and abrupt spatial changes. Third, simulation accuracy still has room for improvement. The model shows limited capacity to capture fine-grained efficiency changes inside urban areas. Regions with strong spatial heterogeneity include urban-rural fringe zones and ecologically sensitive areas. Simulation outcomes in these regions remain to be optimized. Fourth, the operational efficiency of the model needs reasonable upgrading. Data processing and model training for large-scale and long time series require high-performance hardware. Continuous expansion of research data scale will restrict operating speed and overall computational efficiency.

In view of the above limitations, future research can focus on four major directions. First, research can deepen the existing model coupling mechanism. New integration methods can strengthen coordination and interaction among all sub-modules. Improved linkage will enhance the comprehensive capacity to identify temporal and spatial features and evolutionary rules of TSUE. Second, follow-up studies can optimize the system of driving factors. New atypical influencing factors can supplement existing indicator systems. The establishment of a dynamic factor database can strengthen model response to abrupt changes. Such improvement can also promote the comprehensiveness of long-term prediction results. Third, technical upgrading can balance simulation accuracy and operational efficiency. Advanced deep learning algorithms include Transformer and Graph CNN models. These methods can optimize modules for temporal and spatial feature extraction. Refined programming logic and standardized data processing workflows can reduce hardware dependence during model operation. Fourth, subsequent research can expand the application scope of the extended CA model. Close integration with practical demands helps connect model simulation with territorial spatial planning, ecological protection, and regional development. Multi-scenario simulation and conditional prediction can raise practical application value of relevant methods. Relevant research outcomes can provide reliable decision support for the high-quality use of territorial space.

Competing Interests The authors have no conflicts of interest to declare that are relevant to the content of this chapter.

References

1. Sarkar P (2000) A brief history of cellular automata. ACM Comput Surv 32(1):80–107
2. Wolfram S (1983) Statistical-mechanics of cellular automata. Rev Mod Phys 55(3):601–644
3. Uddin MS, Mahalder B, Mahalder D (2023) Assessment of land use land cover changes and future predictions using CA-ANN simulation for Gazipur City Corporation. Bangladesh Sustain 15(16):12329
4. Hajjar SE, Abdallah F, Kassem H et al (2023) Long short-term memory and attention models for simulating urban densification. Sustain Cities Soc 98:104852
5. Zhang P, Liu L, Yang LW et al (2023) Exploring the response of ecosystem service value to land use changes under multiple scenarios coupling a mixed-cell cellular automata model and system dynamics model in Xi'an. China Ecol Indic 147:110009

6. Al-Sahaf H, Bi Y, Chen Q et al (2019) A survey on evolutionary machine learning. J R Soc N Z 49(2):205–228
7. Biamonte J, Wittek P, Pancotti N et al (2017) Quantum machine learning. Nature 549(7671):195–202
8. Guan W, Perdue G, Pesah A et al (2021) Quantum machine learning in high energy physics. Mach Learn-Sci Technol 2(1):011003
9. Greener JG, Kandathil SM, Moffat L et al (2022) A guide to machine learning for biologists. Nat Rev Mol Cell Biol 23(1):40–55
10. Hubel DH (1985) Citation classic-receptive-fields, binocular interaction and functional architecture in the cat's visual cortex. Curr Contents/Life Sci 19:23–23
11. Gu JX, Wang ZH, Kuen J et al (2018) Recent advances in convolutional neural networks. Pattern Recogn 77:354–377
12. Lecun Y, Bottou L, Bengio Y et al (1998) Gradient-based learning applied to document recognition. Proc IEEE 86(11):2278–2324
13. Ghaderizadeh S, Abbasi-Moghadam D, Sharifi A et al (2021) Hyperspectral image classification using a hybrid 3d-2d convolutional neural networks. IEEE J Sel Top Appl Earth Obs Remote Sens 14:7570–7588
14. Kattenborn T, Leitloff J, Schiefer F et al (2021) Review on convolutional neural networks (CNN) in vegetation remote sensing. ISPRS J Photogramm Remote Sens 173:24–49
15. Sun YA, Xue B, Zhang MJ et al (2020) Evolving deep convolutional neural networks for image classification. IEEE Trans Evol Comput 24(2):394–407
16. Kruthiventi SSS, Ayush K, Babu RV (2017) DeepFix: a fully convolutional neural network for predicting human eye fixations. IEEE Trans Image Process 26(9):4446–4456
17. Wieczorek M, Silka J, Wozniak M et al (2022) Lightweight convolutional neural network model for human face detection in risk situations. IEEE Trans Industr Inform 18(7):4820–4829
18. Hochreiter S, Schmidhuber J (1997) Long short-term memory. Neural Comput 9(8):1735–1780
19. Sherstinsky A (2020) Fundamentals of recurrent neural network (RNN) and long short-term memory (LSTM) network. Phys D-Nonlinear Phenom 404:132306
20. Yan R, Liao JQ, Yang J et al (2021) Multi-hour and multi-site air quality index forecasting in Beijing using CNN, LSTM, CNN-LSTM, and spatiotemporal clustering. Expert Syst Appl 169:114513
21. Li X, Peng L, Yao XJ et al (2017) Long short-term memory neural network for air pollutant concentration predictions: method development and evaluation. Environ Pollut 231:997–1004
22. Zheng HF, Lin F, Feng XX et al (2021) A hybrid deep learning model with attention-based conv-LSTM networks for short-term traffic flow prediction. IEEE Trans Intell Transp Syst 22(11):6910–6920
23. Kim HY, Won CH (2018) Forecasting the volatility of stock price index: A hybrid model integrating LSTM with multiple GARCH-type models. Expert Syst Appl 103:25–37
24. Antoniadis A, Lambert-Lacroix S, Poggi JM (2021) Random forests for global sensitivity analysis: a selective review. Reliab Eng & Syst Saf 206:107312
25. Breiman L (2001) Random forests. Mach Learn 45(1):5–32
26. Belgiu M, Dragut L (2016) Random Forest in remote sensing: a review of applications and future directions. ISPRS J Photogramm Remote Sens 114:24–31
27. Sheykhmousa M, Mahdianpari M, Ghanbari H et al (2020) Support vector machine versus random forest for remote sensing image classification: a meta-analysis and systematic review. IEEE J Sel Top Appl Earth Obs Remote Sens 13:6308–6325
28. Speiser JL, Miller ME, Tooze J et al (2019) A comparison of random forest variable selection methods for classification prediction modeling. Expert Syst Appl 134:93–101
29. Stafoggia M, Bellander T, Bucci S et al (2019) Estimation of daily PM10 and PM2.5 concentrations in Italy, 2013-2015, using a spatiotemporal land-use random-forest model. Environ Int 124:170–179
30. Georganos S, Grippa T, Vanhuysse S et al (2018) Less is more: Optimizing classification performance through feature selection in a very-high-resolution remote sensing object-based urban application. Giscience & Remote Sens 55(3):1408892

31. Dong YH, Zhao G, Ma YJ et al (2022) A novel image encryption scheme based on pseudo-random coupled map lattices with hybrid elementary cellular automata. Inf Sci 593:121–154
32. Liu XP, Ma L, Li X et al (2014) Simulating urban growth by integrating landscape expansion index (LEI) and cellular automata. Int J Geogr Inf Sci 28(1):148–163
33. Vispoel M, Daly AJ, Baetens JM (2022) Progress, gaps and obstacles in the classification of cellular automata. Phys D-Nonlinear Phenom 432:133074
34. Lukas P, Melesse AM, Kenea TT (2023) Prediction of future land use/land cover changes using a coupled CA-ANN model in the upper Omo-Gibe River Basin. Ethiop Remote Sens 15(4):1148
35. Silva EA, Ahern J, Wileden J (2008) Strategies for landscape ecology: An application using cellular automata models. Prog Plan 70:133–177
36. Talukdar S, Singha P, Mahato S et al (2020) Land-use land-cover classification by machine learning classifiers for satellite observations-A review. Remote Sens 12(7):1135
37. Liu MH, Chen HY, Qi LI et al (2023) LUCC simulation based on RF-CNN-LSTM-CA model with high-quality seed selection iterative algorithm. Appl Sci-Basel 13(6):3407

Chapter 6
Empirical Analysis: Multi-dimensional Evaluation and Spatiotemporal Variation of Territorial Space Use Efficiency in Chinese Cities

Abstract This chapter measures the TSUE of 340 Chinese cities from 1990 to 2020 based on PLES theory and integrates the GeoFA, GeoDEA, and extended CA models. It establishes a complete evaluation indicator system to calculate PSUE, LSUE, and ESUE, then explores the spatiotemporal evolution and spatial autocorrelation of the three segmented efficiencies as well as overall TSUE, and further predicts their changing trends by 2030. Empirical results indicate that PSUE, LSUE, ESUE, and overall efficiency present differentiated temporal variations with 2010 as the critical turning year, accompanied by prominent spatial imbalance and obvious positive spatial agglomeration. The prediction demonstrates that overall TSUE will achieve steady growth before 2030, while intercity spatial disequilibrium cannot be eliminated thoroughly.

6.1 Research Basis: Study Area, Research Data, and Analytical Framework

6.1.1 Overview of the Study Area

The study adopts a multi-year time series covering 1990, 2000, 2010, 2020, and 2030. Xizang Autonomous Region, Taiwan Province, Hainan Province, Macao Special Administrative Region, and Hong Kong Special Administrative Region show obvious differences from other provincial-level regions in data compilation and release. These differences arise from distinctive historical backgrounds and economic and cultural conditions. Long-term series data for these five regions are difficult to obtain. The research units exclude municipal areas of the above five regions. This exclusion ensures the accuracy and persuasiveness of research results. A total of 340 municipal-level units are included for the evaluation and prediction of TSUE. These units consist of 4 municipalities directly under the central government, 15 sub-provincial cities, 268 prefecture-level cities, 30 autonomous prefectures, 6 prefectures, 14 county-level cities directly under provincial governance, and 3 leagues. The data on boundary scope, place names, and administrative levels reflect the status as of 2019. All

Y. Ma, *Evaluation and Prediction Modeling of Territorial Space Use Efficiency*, Urban Sustainability, https://doi.org/10.1007/978-981-92-4396-9_6

Fig. 6.1 Spatial distribution of the study area[1]

geographic data are derived from the 1:1,000,000 National Fundamental Geographic Database. The data are released on the official website of the National Geomatics Center of China at http://www.ngcc.cn/ngcc/.

As shown in Fig. 6.1, the study area covers a vast land area. The longitudinal range extends from 73°33′E to 135°05′E. The latitudinal range spans from 20°N to 53°33′N. The overall terrain presents a descending trend from the west to the east. The whole region shows typical topographic features of three obvious ladders. Regional climate presents regular zonal distribution from south to north. Main climate types include tropical monsoon climate, subtropical monsoon climate, temperate monsoon climate, and temperate continental climate. The Qinghai-Tibet Plateau and the Altai Mountains are characterized by unique plateau and alpine climate. The study area contains 171 types of mineral resources. Such resources show diverse categories and abundant reserves. Forest resources, plant resources, animal resources, and water resources also reach high levels across the globe. Complex geographical conditions and sufficient natural resources provide solid foundations for rapid social and economic development of cities. The vast spatial span of the study area leads to prominent differences in terrain, climate, and resource endowments. Such disparities drive cities within the study area to adopt differentiated development paths.

[1] This map is compiled based on the base map with the drawing approval number GS (2019) 1822. No modification has been made to the original base map.

China has established itself as the world's second-largest economy. Statistics from authoritative departments show that the national gross domestic product reached 126.06 trillion RMB in 2023. The figure represents an annual growth rate of 5.2% compared with 2022. Agriculture, forestry, manufacturing, construction, and service industries fuel robust growth and global influence. High-tech and innovative industries also maintain outstanding competitiveness across the world. The study area features a large population base. The total population stood at 1.41 billion by the end of 2022 with a stable growth momentum. Population distribution follows a notable spatial pattern. Residents are densely concentrated in the southeast and sparsely distributed in the northwest. Urban population accounts for 66.16% of the national total, while rural population makes up 33.84%. Cities are classified into different grades based on urban population, including megacities, super-large, and large cities. Cities at different levels vary greatly in spatial structure, infrastructure layout, social development, and governance performance. This study carries out empirical research on TSUE. The analysis takes full account of the heterogeneous conditions of 340 cities nationwide. Major influencing factors cover geographical environment, natural resource endowment, and socio-economic development.

6.1.2 Data Sources and Preprocessing

As shown in Table 6.1, this study collects multi-source datasets to conduct the evaluation and prediction of TSUE. All applied data fall into three major categories, namely spatial data, statistical data, and reference literature data.

1. **Spatial Data**

This study adopts multiple types of spatial data including vector data, raster data, and POI data. The study area boundary data defines the geographic scope of each city. Such data are downloaded from the official website of the National Geomatics Center of China. The dataset lays a solid foundation for model construction, efficiency evaluation, and predictive analysis. Land use remote sensing monitoring data support the classification of production space, living space, and ecological space. These data also serve as the basic material for landscape index calculation and green infrastructure area statistics. The data are provided by the Resources and Environmental Science Data Center with a spatial resolution of 1 km × 1 km. Landsat series satellite remote sensing images act as critical auxiliary data. These images support the inversion of humidity, greenness, dryness, and heat indexes. The broad coverage of the study area brings massive data volume and complex processing work. The Google Earth Engine platform performed large-scale index inversion and calculation in this research. DEM data, gridded population data and gridded GDP data serve as essential impact factors. Annual average temperature, annual average precipitation, POI data, and traffic network data are also included. $PM_{2.5}$ datasets are integrated into the prediction process as well. The all above elements function as driving factors for the numerical simulation of TSUE.

Table 6.1 Detailed information of research data

Data category	Data name	Data source	Data type	Spatial resolution	Temporal resolution
Spatial Data	Study Area Boundary Data	National Geomatics Center of China (NGCC), http://www.ngcc.cn/ngcc/	Vector Data (Polygon File)	/	2019
	Land Use Remote Sensing Monitoring Data	Resources and Environmental Science Data Center (RESDC), https://www.resdc.cn/	Raster Data	1 km × 1 km	1990, 2000, 2010, 2020
	Remote Sensing Image Data	Google Earth Engine (GEE), Landsat satellite remote sensing image data	Raster Data	30 m × 30 m	1990, 2000, 2010, 2020
	DEM Data	Resources and Environmental Science Data Center (RESDC), https://www.resdc.cn/	Raster Data	1 km × 1 km	2000
	Gridded Population Distribution Data	Resources and Environmental Science Data Center (RESDC), https://www.resdc.cn/	Raster Data	1 km × 1 km	1990, 2000, 2010, 2020
	Gridded GDP Data	Resources and Environmental Science Data Center (RESDC), https://www.resdc.cn/	Raster Data	1 km × 1 km	1992, 2000, 2010, 2019
	Annual Average Temperature	Resources and Environmental Science Data Center (RESDC), https://www.resdc.cn/	Raster Data	1 km × 1 km	1990, 2000, 2010, 2020
	Annual Average Precipitation	Resources and Environmental Science Data Center (RESDC), https://www.resdc.cn/	Raster Data	1 km × 1 km	1990, 2000, 2010, 2020
	POI Data of Hospitals, Schools and Administrative Centers	Amap	Vector Data (Point File)	/	2019
	Traffic Network Data	Open Street Map	Vector Data	/	2018

(continued)

Table 6.1 (continued)

Data category	Data name	Data source	Data type	Spatial resolution	Temporal resolution
	$PM_{2.5}$ Concentration Data	Dataset from previous studies, https://zenodo.org/records/6372847#.YtPqSTckXWI	Raster Data	0.25° × 0.25°	1990, 2000, 2010, 2020
Statistical Data	GDP, population, fixed asset investment, construction land, enrolled students and others	CNKI China Economic and Social Big Data Research Platform, https://data.cnki.net/yearBook?type=type&code=A	/	/	1990, 2000, 2010, 2020
Reference literature data	Carbon emission coefficient, ecosystem service value equivalent, sewage discharge coefficient and others	Compiled and obtained from relevant literature	/	/	/

Spatial data preprocessing is conducted through Python programming and ArcGIS Pro 3.0 software. The preprocessing ensures consistency of reference coordinate systems, spatiotemporal resolution, and regional coverage among all datasets. Key procedures include projection definition, data clipping, reclassification, and resampling. Additional data calculation is performed based on existing spatial datasets to meet research demands. Typical processes involve the calculation of landscape indices, total carbon emissions, total carbon sequestration, and green infrastructure indicators. All these indicators are derived from land use remote sensing monitoring data. Historical POI data and traffic network data for 1990, 2000, and 2010 are unavailable. Data from 2019 and 2018 are adopted as alternative sources for the missing records. This method represents a common processing strategy adopted by scholars in related studies [1].

2. **Statistical Data**

Statistical data in this study are mainly sourced from urban statistical yearbooks. These datasets cover social, economic, demographic, and resource-related conditions with complex and diversified contents. Statistical indicators serve as core measurement factors for TSUE. Selected indicators are also adopted as driving factors for efficiency prediction. Long-term cross-sectional data of 1990, 2000, 2010, and 2020 are required for chronological analysis. Changes in statistical standards, administrative adjustments, and divergent statistical calibers lead to technical defects. Common defects include inconsistent measurement units and partial data missing. Three standardized methods are applied to address the above problems. Relevant data of adjacent years replace indicators without official records in specific periods. Mean value method, exponential smoothing method, and trend analysis supplement missing statistical values. Data merging processing is adopted for regions with administrative division adjustment. Cities with similar development stages and evolutionary patterns provide alternative data for special samples. Basic mathematical calculation is conducted to unify dimension specifications for indicators with inconsistent measurement units.

3. **Literature-Based Parameter Data**

This study involves a wide range of indicator datasets. Several indicators require targeted processing and calculation based on original basic data. The selection of relevant coefficient parameters refers to published research findings. Literature-derived data, therefore, constitute an essential data source for this research. Typical parameters include carbon emission coefficients, ecosystem service value equivalents, and sewage discharge coefficients. Other core indicators cover habitat type weights, chemical fertilizer pollution coefficients, pesticide pollution coefficients, and plastic film pollution coefficients.

6.1.3 Overall Analytical Framework: From Efficiency Measurement to Spatiotemporal Analysis and Prediction

Supported by multi-source data, this study analyzes the structure and functions of the territorial space use system. It constructs an evaluation and prediction framework for TSUE based on the PLES use system. The overall framework is presented in Fig. 6.2.

Firstly, this study defines the research orientation and completes preliminary preparation. This section contains two core modules of policy support and multi-source data support. The first module takes the sustainable development of territorial space use as the overall goal. It reviews national strategic documents such as ecological civilization construction, territorial spatial master planning, multi-plan integration, and unified spatial information platform. The review clarifies existing problems and governance directions of TSUE, and establishes a solid policy and theoretical foundation. The second module sorts out the heterogeneous relations of multi-source data including statistical data, survey data, remote sensing data, and measurement data. Standardized preprocessing is implemented to unify temporal scales, spatial scales and research scopes, so as to lay a reliable data foundation for subsequent analysis.

Secondly, this study constructs evaluation and prediction models to form methodological support. The model system consists of the GeoFA, the GeoDEA and the extended CA models. These tools are separately applied to indicator system establishment, quantitative measurement, and dynamic prediction of TSUE. As shown in Fig. 6.2, a scientifically constructed indicator system acts as the basic premise of the

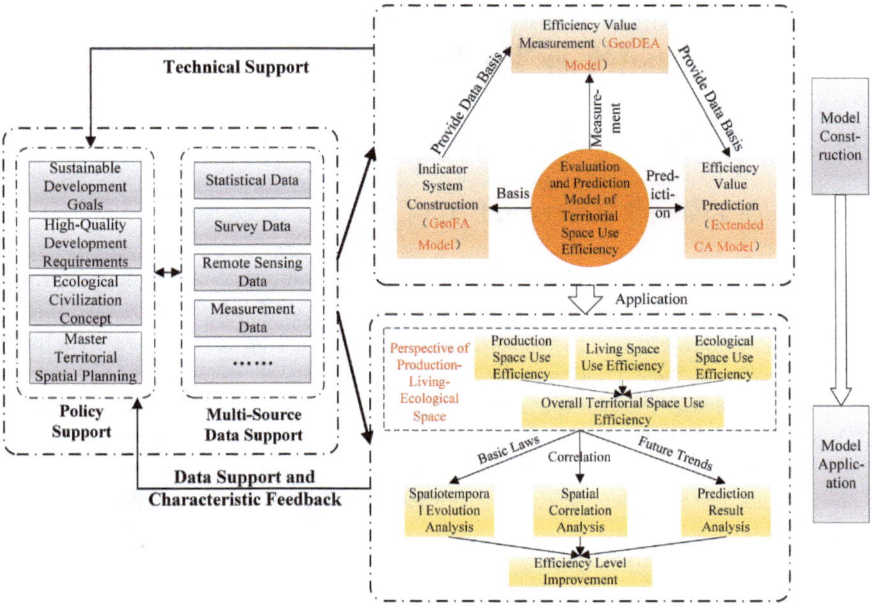

Fig. 6.2 Research framework of territorial space use efficiency

whole research and provides clear guidance and data support for efficiency measurement. Efficiency measurement serves as a critical intermediate link in the overall research process. Relevant indicators are imported to calculate efficiency values of historical and current periods. Measured results are further adopted as input parameters for simulation prediction, so as to realize the dynamic prediction of TSUE. This research eventually forms a complete research paradigm for efficiency evaluation and prediction and offers technical reference for the long-term monitoring of territorial space use.

Finally, empirical analysis verifies the rationality and applicability of the constructed models. Corresponding conclusions are proposed to promote the overall improvement of national TSUE. On the basis of the above theoretical basis and technical methods, this study selects 340 municipal units across China as the research cases. It completes the indicator system construction, quantitative measurement, and simulation prediction for PSUE, LSUE, ESUE, and overall TSUE. The research further explores spatiotemporal evolution characteristics and spatial correlation features of efficiency values. It explains prediction results in detail and proposes targeted policy suggestions for efficiency optimization.

6.2 Construction of Evaluation Indicator System for Territorial Space Use Efficiency

6.2.1 Construction of PSUE Indicator System

6.2.1.1 Preliminary Screening of PSUE Indicators

Measurement indicators of PSUE evaluate the input–output relationship of various production activities. These indicators cover industry, agriculture, forestry, animal husbandry, fishery, and other production sectors. Production activities rely on basic elements such as material resources, labor force, capital investment, and spatial patterns. This study selects initial input indicators from four dimensions of resource, labor, capital, and landscape. The resource dimension includes natural resources and non-natural resources. Natural resources reflect geographical spatial elements and highlight the unique characteristics of territorial space. The landscape dimension focuses on spatial patterns from the perspective of land patch granularity in territorial space use. Output indicators take both socio-economic benefits and negative ecological and landscape impacts into consideration. Initial output indicators are thus determined from four dimensions of economy, society, landscape, and pollutant emission. As shown in Table 6.2, a total of 39 indicators are initially selected for PSUE measurement. The indicator system contains 17 input indicators and 22 output indicators.

The sources and calculation methods of all indicators are specified in Table 6.2. This study omits detailed descriptions for simple calculation processes. The calculation formulas of major indicators are illustrated in the following content.

1. **Calculation of Production Space Score**

Referring to existing research findings, this study takes the correspondence diagram between PLES system and land use types as the basic premise [2], as presented in Fig. 1.3. The production space score is obtained by assigning graded scores to different land use types, with specific grading standards listed in Table 6.3. The scoring system is divided into four grades according to functional intensity. The highest score is set at 5 points, the medium level at 3 points, and the lowest level at 1 point. Land types without corresponding production function are marked with 0 point. This study adopts China's land use remote sensing monitoring data provided by the Institute of Geographic Sciences and Natural Resources Research, Chinese Academy of Sciences as the basic dataset. The final score is calculated by summing the product of the area and corresponding score of each land use type.

2. **Calculation of Total Agricultural Resource Input**

In this study, the total agricultural resource input is defined as the sum of three indicators, namely the consumption of agricultural chemical fertilizers, agricultural

Table 6.2 Preliminary indicators for measurement of production space use efficiency

Indicator type	Indicator dimension	Specific indicator	Abbreviation	Indicator description	Indicator source and calculation method
Input Indicators	Resource Input (Natural Resources, Non-natural Resources)	Mean DEM (m)	In_DEM	The average DEM value within the DMU	Calculated independently based on DEM raster data
		Production Space Score	In_ScVa	The evaluation score of production space within the DMU	Calculated independently (see the following content)
		Cultivated Land Area (km^2)	In_CuAr	The cultivated land area within the DMU	Statistically calculated independently based on land use data
		Total Gas Supply (Coal Gas, Natural Gas) (10^4 m^3)	In_GaSu	The total supply of coal gas and natural gas within the DMU	Statistical data
		Total Liquefied Petroleum Gas Supply (t)	In_LiPe	The total supply of liquefied petroleum gas within the DMU	Statistical data
		Urban Road Area at Year-end (10^4 m^2)	In_RoAr	The actual urban road area at year-end within the DMU	Statistical data
		Total Agricultural Resource Input (t)	In_AgIn	The total consumption of agricultural chemical fertilizers, plastic films and pesticides during the research period	Calculated independently from statistical data (see the following content)
		Number of Industrial Enterprises	In_EnNu	The total number of industrial enterprises within the DMU	Statistical data
		Built-up Area (km^2)	In_BuAr	The total built-up area within the DMU	Statistical data

(continued)

Table 6.2 (continued)

Indicator type	Indicator dimension	Specific indicator	Abbreviation	Indicator description	Indicator source and calculation method
	Labor Input	Final Employees of Urban Non-private Units (person)	In_EmNu	The total number of employees in urban non-private units at the end of the research period	Statistical data
		Average On-the-job Workers of Urban Non-private Units (10^4 persons)	In_EmAnu	The average number of on-the-job workers in urban non-private units during the research period	Statistical data
		Total Population (10^4 persons)	In_Popu	The total population within the DMU during the research period	Statistical data
	Capital Input	Fixed Asset Investment (10^4 yuan)	In_FiIn	The total fixed asset investment within the DMU during the research period	Statistical data
		Local General Public Budget Expenditure (10^4 yuan)	In_BuEx	The total local general public budget expenditure within the DMU during the research period	Statistical data
	Landscape Input	Number of Patches	In_NP	The number of landscape patches within the DMU	Calculated independently based on land use data
		Patch Density	In_PD	The landscape patch density within the DMU	Calculated independently based on land use data
		Landscape Shape Index	In_LSI	The value of landscape shape index within the DMU	Calculated independently based on land use data

(continued)

Table 6.2 (continued)

Indicator type	Indicator dimension	Specific indicator	Abbreviation	Indicator description	Indicator source and calculation method
Output Indicators	Economic Output	Regional Gross Domestic Product (Current Price) (10^8 yuan)	Out_GDP	The regional gross domestic product within the DMU	Calculated independently based on land use data
		Per Capita Regional GDP (yuan)	Out_RGDP	The ratio of regional gross domestic product to total population within the DMU	Calculated independently from statistical data
		Regional GDP Growth Rate (%)	Out_GDPr	The growth rate of regional gross domestic product compared with the previous period	Calculated independently from statistical data
		Proportion of Primary Industry Output in Regional GDP (%)	Out_PrPs	The proportion of primary industry output in regional gross domestic product	Calculated independently from statistical data
		Proportion of Secondary Industry Output in Regional GDP (%)	Out_PrSs	The proportion of secondary industry output in regional gross domestic product	Calculated independently from statistical data
		Proportion of Tertiary Industry Output in Regional GDP (%)	Out_PrTs	The proportion of tertiary industry output in regional gross domestic product	Calculated independently from statistical data
		Assets and Profits of Industrial Enterprises Above Designated Size (10^4 yuan)	Out_EnPr	The total assets and total profits of industrial enterprises above designated size	Statistical data

(continued)

Table 6.2 (continued)

Indicator type	Indicator dimension	Specific indicator	Abbreviation	Indicator description	Indicator source and calculation method
		Local General Public Budget Revenue (10^4 yuan)	Out_BuIn	The local general public budget revenue within the DMU	Statistical data
	Social Output	Total Retail Sales of Consumer Goods (10^4 yuan)	Out_Trscc	The total retail sales of consumer goods within the DMU	Statistical data
		Total Wages of On-the-job Workers in Urban Non-private Units (10^4 yuan)	Out_ToWa	The total wages of on-the-job workers in urban non-private units	Statistical data
		Average Wage of On-the-job Workers in Urban Non-private Units (yuan)	Out_ToRwa	The average wage of on-the-job workers in urban non-private units	Statistical data
	Landscape Output	Contagion Index	Out_CONT	The contagion index within the DMU	Calculated independently based on land use data
		Landscape Division Index	Out_DIVI	The landscape division index within the DMU	Calculated independently based on land use data
		Effective Mesh Size	Out_MESH	The effective mesh size within the DMU	Calculated independently based on land use data
		Splitting Index	Out_SPLI	The splitting index within the DMU	Calculated independently based on land use data
		Patch Richness Density	Out_PRD	The patch richness density within the DMU	Calculated independently based on land use data

(continued)

Table 6.2 (continued)

Indicator type	Indicator dimension	Specific indicator	Abbreviation	Indicator description	Indicator source and calculation method
		Shannon Diversity Index	Out_SHDI	The Shannon diversity index within the DMU	Calculated independently based on land use data
		Shannon Evenness Index	Out_SHEI	The Shannon evenness index within the DMU	Calculated independently based on land use data
		Simpson Evenness Index	Out_SIEI	The Simpson evenness index within the DMU	Calculated independently based on land use data
	Pollutant Output	Total Agricultural Pollution (t)	Out_AgPo	The total pollution load caused by chemical fertilizers, pesticides and plastic films	Calculated independently from statistical data (see the following)
		Total Carbon Emissions (t)	Out_CaEm	The total carbon emissions within the DMU	Calculated independently from statistical data (see the following)
		Industrial Particulate Matter Emissions (t)	Out_IpEm	The industrial particulate matter emission within the DMU	Statistical data

plastic films, and pesticides. Relevant data collected from statistical yearbooks are based on provincial administrative units. Accordingly, the input of each agricultural resource at the municipal scale is calculated as Eq. (6.1), and the total agricultural resource input is obtained by cumulative summation.

$$\begin{aligned}&\text{Municipal agricultural resource input}\\&\quad = \text{Provincial agricultural resource input/Total provincial cultivated land area}\\&\qquad \times \text{Municipal cultivated land area}\end{aligned} \tag{6.1}$$

Table 6.3 Scoring assignment criteria for production-living-ecological space

First-class land use classification		Second-class land use classification		Production space score per square kilometer/km^2	Living space score/km^2	Ecological space score/km^2
Code	Name	Code	Name			
1	Cultivated Land	11	Paddy field	3	0	3
		12	Dry farmland	3	0	3
2	Forest Land	21	Arbour forest	0	0	5
		22	Shrub forest	0	0	5
		23	Open forest	0	0	5
		24	Other forest land	0	0	5
3	Grassland	31	High-coverage grassland	1	0	5
		32	Medium-coverage grassland	1	0	5
		33	Low-coverage grassland	1	0	5
4	Water Body	41	Rivers and canals	1	0	5
		42	Lakes	1	0	5
		43	Reservoirs and ponds	1	0	5
		44	Permanent glaciers and snowfields	0	0	5
		45	Tidal flats	0	0	5
		46	Floodplain land	0	0	5
5	Urban, Rural, Industrial, and Residential Land	51	Urban construction land	5	3	0
		52	Rural residential areas	0	5	0
		53	Other construction land	5	0	0
6	Unused Land	61	Sandy land	0	0	5
		62	Gobi desert	0	0	5
		63	Saline-alkali land	0	0	5
		64	Wetland marshes	0	0	5
		65	Bare land	0	0	5
		66	Bare rocky land	0	0	5
		67	Other lands	0	0	5
9	Ocean Area	99	Ocean area	0	0	5

Table 6.4 Carbon emission coefficients of different land use types

Land use type	Carbon emission coefficient ($t/hm^2 \cdot a$)
Cultivated land	0.422
Forest land	−0.644
Grassland	−0.021
Water area	−0.253
Urban, rural, industrial, and residential land	42.970
Unused land	−0.005

3. **Calculation of Landscape Metrics**

All landscape metrics in this study are calculated by Fragstats 4.2 software.

4. **Calculation of Total Agricultural Pollution**

Total agricultural pollution covers three pollution sources. These sources include chemical fertilizer pollution, plastic film pollution, and pesticide residue pollution. The corresponding conversion ratios are set separately. The ratio for chemical fertilizer input is 65%. The ratio for plastic film input is 10%. The ratio for pesticide input is 50%. The original input data are calculated in accordance with Eq. (6.1).

5. **Calculation of Total Carbon Emissions**

This study fully considers research objectives, theoretical basis, and core themes. The measurement of total carbon emissions adopts the land use based emission coefficient method from the perspective of land science. This method calculates total carbon emissions through cumulative multiplication. The calculation multiplies the area of each land use type by its matched carbon emission coefficient. Carbon emission coefficients in this study refer to existing research results and national domestic conditions of China [3–5]. Specific coefficient values are listed in Table 6.4.

6.2.1.2 PSUE Indicator Determination via GeoFA Model

1. **Input Indicators**

This study adopts seventeen preliminary input indicators mentioned above, as presented in Table 6.2. The GeoFA model generates the common factor loading matrix through operational calculation, as illustrated in Fig. 6.3. Indicators from the resource input dimension show distinct high loading values for cultivated land area (In_CuAr), built-up area (In_BuAr), and total gas supply including coal gas and natural gas (In_GaSu). In_CuAr records the highest loading value in Common Factor 3 of 1990, 2000, and 2020, as well as Common Factor 2 of 2010. In_BuAr ranks first in Common Factor 4 of 2000 and Common Factor 1 of 2010. This indicator also

Fig. 6.3 Heatmap of factor loading matrix for input indicators of production space use efficiency[2]

maintains a high loading value and ranks third in Common Factor 1 of 2020. In_GaSu takes the leading position in loading values in Common Factor 4 of 1990, Common Factor 5 of 2000, and Common Factor 6 of 2010. Two labor input indicators reflect prominent loading performances across common factors. These indicators refer to the number of staff in urban non-private units at the year-end (In_EmNu) and the average number of on-the-job employees in urban non-private units (In_EmAnu). The loading value of In_EmAnu reaches 0.91 in Common Factor 6 of 2010. This loading value within the corresponding common factor shows an obvious numerical gap with other indicators. Fixed asset investment (In_FiIn) serves as the core capital input indicator with high loading values in multiple common factors. This indicator ranks first in Common Factor 1 of 2000 and Common Factor 4 of 2020. In the landscape dimension, the landscape shape index (In_LSI) exhibits higher loading values than the number of patches (In_NP) and patch density (In_PD). This study conducts comprehensive screening on the basis of the above analysis. Six indicators are finally determined as the input indicators for PSUE, namely In_BuAr, In_CuAr, In_GaSu, In_EmAnu, In_FiIn, and In_LSI.

[2] The full definitions of abbreviated indicators in the figure are listed in Table 6.2.

2. **Output Indicators**

The GeoFA model is applied to analyze twenty-two preliminary output indicators of PSUE (see Table 6.2), and the corresponding common factor loading matrix is presented in Fig. 6.4. Multiple indicators reflecting economic output present high loading values in common factors. Economic indicators thus act as core metrics for evaluating PSUE. Three indicators obtain the highest loading values in numerous common factors. These indicators include regional gross domestic product (Out_GDP), regional GDP growth rate (Out_GDPr), and the proportion of the secondary industry in regional GDP (Out_PrSs). Relevant common factors cover Factor 2, Factor 3 and Factor 6 in 1990, Factor 4 and Factor 6 in 2000, as well as Factor 1, Factor 3 and Factor 6 in 2010 and 2020. Social output indicators also show considerable loading values in common factors, while relevant values remain lower than those of economic output indicators. Output indicators from the landscape dimension occupy an essential position in factor loading results. Shannon's diversity index (Out_SHDI) and patch richness density (Out_PRD) rank first in the loading values of multiple common factors. Out_PRD maintains high loading values in the common factors of 1990, 2000, and 2020. This indicator shows prominent numerical gaps with other indicators and serves as the dominant factor. From the perspective of pollutant emission, industrial particulate matter emission (Out_IpEm) exhibits prominent factor loadings in the common factors of all study years with obvious gaps from other indicators. This result demonstrates the critical role of this indicator. After comprehensive comparison and analysis, six indicators are determined as the final output indicators of PSUE. The selected indicators consist of Out_GDP, Out_GDPr, Out_PrSs, Out_PRD, Out_SHDI and Out_IpEm. The first four items are the desirable outputs, and the last two items are classified as undesirable outputs.

6.2.2 *Construction of LSUE Indicator System*

6.2.2.1 Preliminary Screening of LSUE Indicators

LSUE mainly reflects the quality of living space. It fully covers the space required for various daily human activities, including residential space, employment space, educational space, medical and health space, recreation space, and other functional spaces. The preliminary selection of measurement indicators for LSUE focuses on the core function of living space. Such function refers to diverse public service supplies for daily life activities. Consistent with the selection logic of PSUE indicators, this study incorporates natural resources, landscape patches, and pollutant conditions into the analytical framework. These dimensions can reflect the supporting effects of land resources, energy resources, and ecological environment on living space. The adverse social and environmental impacts caused by regular human life activities

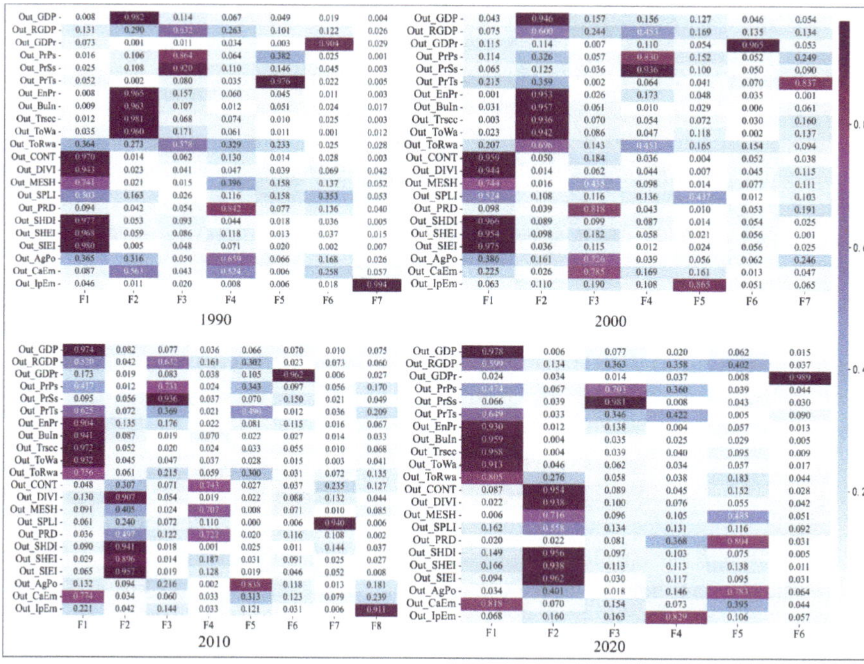

Fig. 6.4 Heatmap of factor loading matrix for output indicators of production space use efficiency[3]

are also taken into account. The preliminary input indicators are classified into five dimensions, namely resource supply, medical care, education, culture, and capital investment. The preliminary output indicators cover seven dimensions involving economy, society, population, landscape, transportation, recreation, and pollutant emission. As shown in Table 6.5, a total of 38 indicators are initially selected for measuring LSUE, consisting of 18 input indicators and 20 output indicators.

The sources and calculation methods of all indicators are specified in Table 6.5. For indicators with straightforward calculation procedures, detailed descriptions are omitted in this study. The calculation methods of partial indicators are presented as follows.

1. **Calculation of Green Infrastructure Area**

In the preliminary exploration stage of this study, substantial experimental foundations and research achievements have been accumulated regarding the calculation of green infrastructure area. The specific calculation method refers to the published literature of Ma et al. [6, 7].

[3] The full definitions of abbreviated indicators in the figure are listed in Table 6.2.

Table 6.5 Preliminary indicators for measuring living space use efficiency

Indicator type	Indicator perspective	Specific indicator	Abbreviation	Indicator description	Indicator source and calculation method
Input Indicators	Resource Input (Natural and Non-natural Resources)	Mean DEM (m)	In_DEM	Mean DEM value within the DMU	Calculated independently based on DEM raster data
		Living Space Score	In_ShVa	Living space evaluation score within the DMU	Calculated independently (refer to Sect. 6.2.1.1)
		Total Household Gas Supply (Coal Gas, Natural Gas) (10^4 m^3)	In_FgaSu	Total household supply of coal gas and natural gas within the DMU	Statistical data
		Total Household Liquefied Petroleum Gas Supply (t)	In_FliPe	Total household supply of liquefied petroleum gas within the DMU	Statistical data
		Built-up Area (km^2)	In_BuAr	Total built-up area within the DMU	Statistical data
		Core Area of Green Infrastructure (km^2)	In_GiAr	Total area of green infrastructure within the DMU	Calculated independently based on land use data (see below)
	Medical and Health Input	Number of Hospitals	In_HoAu	Number of hospitals within the DMU	Statistical data
		Number of Hospital Beds	In_HoBe	Number of hospital beds within the DMU	Statistical data
		Number of Licensed Physicians	In_DoNu	Number of licensed physicians within the DMU	Statistical data
	Education Input	Number of Regular Institutions of Higher Education	In_HedNu	Number of regular institutions of higher education within the DMU	Statistical data

(continued)

Table 6.5 (continued)

Indicator type	Indicator perspective	Specific indicator	Abbreviation	Indicator description	Indicator source and calculation method
		Number of Regular Middle Schools	In_MedNu	Number of regular middle schools within the DMU	Statistical data
		Number of Regular Primary Schools	In_PedNu	Number of regular primary schools within the DMU	Statistical data
		Number of Full-time Teachers in Regular Institutions of Higher Education	In_HteNu	Number of full-time teachers in regular institutions of higher education within the DMU	Statistical data
		Number of Full-time Teachers in Regular Middle Schools	In_MteNu	Number of full-time teachers in regular middle schools within the DMU	Statistical data
		Number of Full-time Teachers in Regular Primary Schools	In_PteNu	Number of full-time teachers in regular primary schools within the DMU	Statistical data
	Cultural Input	Number of Cultural Venues	In_CuNu	Number of museums or cinemas within the DMU	Statistical data
	Capital Input	Expenditure on Science and Technology (10^4 Yuan)	In_TeEx	Total investment in science and technology within the DMU during the research period	Statistical data
		Education Expenditure (10^4 Yuan)	In_EdEx	Total investment in education within the DMU during the research period	Statistical data

(continued)

Table 6.5 (continued)

Indicator type	Indicator perspective	Specific indicator	Abbreviation	Indicator description	Indicator source and calculation method
Output Indicators	Economic Output	Regional Gross Domestic Product (Current Price) (10^8 Yuan)	Out_GDP	Regional gross domestic product within the DMU	Statistical data
		Per Capita Regional GDP (Yuan/person)	Out_RGDP	Ratio of regional gross domestic product to total population within the DMU	Calculated independently from statistical data
		Regional GDP Growth Rate (%)	Out_GDPr	Growth rate of regional gross domestic product compared with the previous period in the research period	Calculated independently from statistical data
	Social Output	Total Retail Sales of Consumer Goods (10^4 Yuan)	Out_Trscc	Total retail sales of consumer goods within the DMU	Statistical data
		Per Capita Retail Sales of Consumer Goods (Yuan/person)	Out_Rtrscc	Ratio of total retail sales of consumer goods to total population within the DMU	Calculated independently from statistical data
		Total Wages of On-the-job Employees in Urban Non-private Units (10^4 Yuan)	Out_ToWa	Total wages of on-the-job employees in urban non-private units within the DMU	Statistical data
		Average Wage of On-the-job Employees in Urban Non-private Units (Yuan)	Out_ToRwa	Average wage of on-the-job employees in urban non-private units within the DMU	Statistical data

(continued)

Table 6.5 (continued)

Indicator type	Indicator perspective	Specific indicator	Abbreviation	Indicator description	Indicator source and calculation method
	Population Output	Total Population (10^4 persons)	Out_Popu	Total population within the DMU during the research period	Statistical data
		Population Density (persons/km^2)	Out_PoDe	Ratio of total population to living space area within the DMU	Calculated independently from statistical data
		Natural Population Growth Rate (‰)	Out_NaGrR	Ratio of natural population growth to annual average population within the DMU during the research period	Statistical data
		Annual Registered Urban Unemployed Population (10^4 persons)	Out_UemNu	Number of registered urban unemployed population at the end of the research period within the DMU	Statistical data
		Year-end Number of Employees in Urban Non-private Units	Out_EmNu	Total number of employees in urban non-private units on the last day of the research period within the DMU	Statistical data
		Number of Students Enrolled	Out_SuNu	Total number of enrolled students in universities, regular middle schools and regular primary schools within the DMU	Calculated independently from statistical data
	Landscape Output	Shannon Diversity Index	Out_SHDI	Shannon diversity index within the DMU	Calculated independently based on land use data

(continued)

Table 6.5 (continued)

Indicator type	Indicator perspective	Specific indicator	Abbreviation	Indicator description	Indicator source and calculation method
		Shannon Evenness Index	Out_SHEI	Shannon evenness index within the DMU	Calculated independently based on land use data
	Transportation Output	Total Passenger Traffic (10^4 persons)	Out_PaTo	Total passenger traffic of railway, highway, waterway, and civil aviation within the research period	Statistical data
		Total Freight Traffic (t)	Out_FrTo	Total freight traffic of railway, highway, waterway, and civil aviation within the research period	Statistical data
	Leisure and recreation output	Per Capita Core Area of Green Infrastructure (m^2/person)	Out_GiAar	Ratio of total core area of green infrastructure to total population within the DMU	Calculated independently based on land use data and statistical data
	Pollutant Output	Domestic Sewage Discharge (t)	Out_SeOu	Domestic sewage discharge within the DMU	Calculated independently (see below)
		Total Carbon Emissions (t)	Out_CaEm	Total carbon emissions within the DMU	Calculated independently based on land use data (refer to Sect. 6.2.1.1)

2. **Calculation of Domestic Sewage Discharge**

The domestic sewage discharge is calculated as Eqs. (6.2)–(6.4).

$$\text{Domestic Sewage Discharge} = \text{Urban Domestic Sewage Discharge} + \text{Rural Domestic Sewage Discharge} \quad (6.2)$$

$$\begin{aligned}&\text{Urban Domestic Sewage Discharge } (10^4 \text{ tons})\\&= \text{Urban Permanent Population } (10^4 \text{ persons})\\&\quad\times \text{Per Capita Comprehensive Domestic Water Consumption Coefficient } (L/(\text{person}\cdot\text{day}))\\&\quad\times \text{Sewage Conversion Coefficient} \times 365\,(\text{days})/1000\end{aligned} \tag{6.3}$$

$$\begin{aligned}&\text{Rural Domestic Sewage Discharge } (10^4 \text{ tons})\\&= \text{Rural Permanent Population } (10^4 \text{ persons})\\&\quad\times \text{Sewage Discharge Coefficient } (L/(\text{person}\cdot\text{day}))\\&\quad\times 365\,(\text{days})/1000\end{aligned} \tag{6.4}$$

Among them, the sewage conversion coefficient and sewage discharge coefficient are derived from the Manual of Production and Discharge Coefficients for Domestic Pollution Sources issued by the Ministry of Ecology and Environment of the People's Republic of China.

6.2.2.2 LSUE Indicator Calibration via GeoFA Model

1. **Input Indicators**

The GeoFA model is adopted to analyze the eighteen initially selected input indicators of LSUE (see Table 6.5), and the common factor loading matrix is obtained, as shown in Fig. 6.5. From the resource perspective, two indicators including the core area of green infrastructure (In_GiAr) and mean DEM value (In_DEM) presented high loading values in the same common factor, namely Common Factor 2 in 1990 and Common Factor 3 in 2000, 2010, and 2020. By comparison, the loading value of In_GiAr was higher than that of In_DEM. In addition, the total household liquefied petroleum gas supply (In_FliPe) ranks highest in corresponding factor loadings of Common Factor 5 in 2000 and 2010, as well as Common Factor 4 in 2020. Among the three initial indicators from the medical and health perspective, only the number of licensed physicians (In_DoNu) showed a relatively high loading value in Common Factor 1 of 1990, while the other two indicators had no prominent performance in the common factors. Indicators from the education perspective had high loading values in multiple common factors with small differences among individual loadings. From the dimensions of school scale and teacher allocation, the number of regular middle schools (In_MedNu) and the number of full-time teachers in regular institutions of higher education (In_HteNu) were finally selected. The number of cultural venues (In_CuNu) exhibits high factor loadings in Common Factor 3 of 1990 and Common Factor 5 of 2000, with distinctly higher loadings than alternative indicators in Common Factor 5 of 2000. From the capital perspective, expenditure on science and technology (In_TeEx) ranked among the top loading values in Common Factor 1 of 1990 and 2010, and Common Factor 4 of 2020. Combined with the actual availability and characteristics of the original data, seven indicators including In_GiAr,

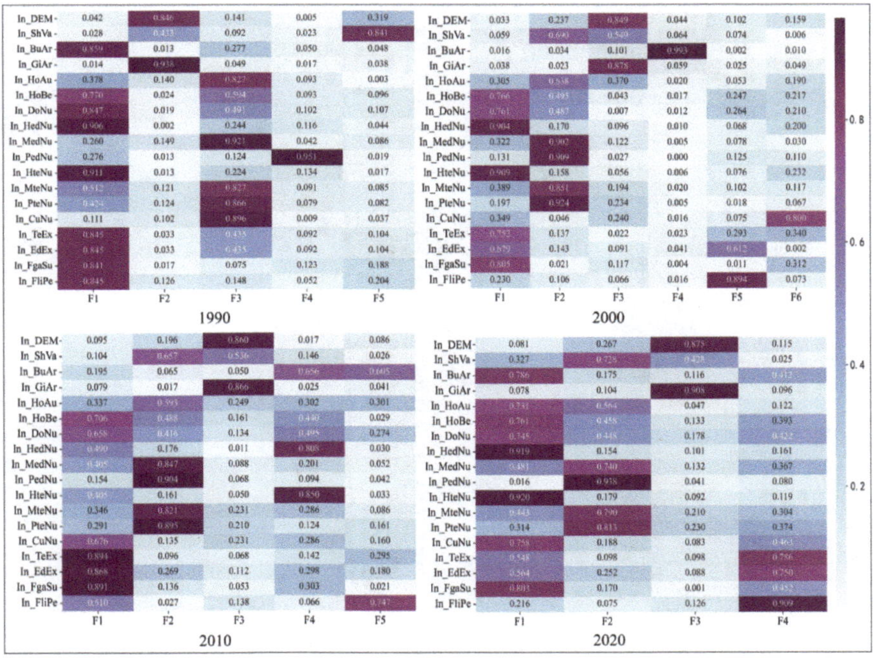

Fig. 6.5 Heatmap of factor loading matrix for input indicators of living space use efficiency[4]

In_FliPe, In_DoNu, In_MedNu, In_HteNu, In_CuNu, and In_TeEx were ultimately determined as the input indicators for LSUE.

2. **Output Indicators**

The GeoFA model is adopted to analyze twenty initially selected output indicators of LSUE (see Table 6.5), and the common factor loading matrix was obtained as shown in Fig. 6.6. The three indicators from the economic perspective all present high loading values in the common factors. Among them, per capita regional GDP (Out_RGDP) and regional GDP growth rate (Out_GDPr) both ranked first in loading values across two common factors. The four indicators from the social perspective also exhibited high loading values in the common factors, which verifies that social indicators can well characterize the development status of living space. Total retail sales of consumer goods (Out_Trscc) were selected as the indicator for the social output perspective. From the population perspective, the natural growth rate (Out_NaGrR) ranked first in loading values in Common Factor 4 of 2000 and Common Factor 5 of 2010, with a notable gap compared with the loading values of other indicators. The two landscape perspective indicators maintained high and identical loading values in Common Factor 2 for 1990, 2000, 2010, and 2020. The Shannon diversity index (Out_SHDI) was selected as the representative indicator for the landscape

[4] The specific connotations of the indicator abbreviations in the figure are shown in Table 6.5.

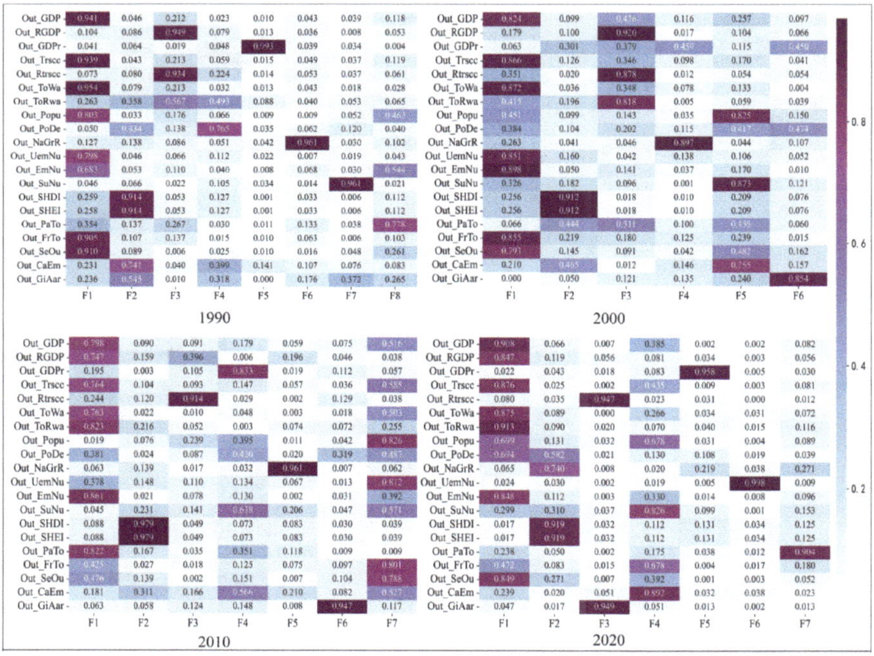

Fig. 6.6 Heatmap of factor loading matrix for output indicators of living space use efficiency[5]

perspective. From the pollutant perspective, domestic sewage discharge (Out_SeOu) showed relatively high loading values in Common Factor 1 of 1990 and 2020. From the leisure and recreation perspective, per capita core area of green infrastructure (Out_GiAar) ranked first in loading values in Common Factor 6 of 2000 and 2020, with a distinct difference from other indicators. After comprehensive consideration, Out_RGDP, Out_Trscc, Out_NaGrR, Out_GiAar, Out_SHDI, and Out_SeOu were finally determined as the output indicators of LSUE. Among them, the first four are desirable output indicators, and the latter two are undesirable output indicators.

6.2.3 Construction of ESUE Indicator System

6.2.3.1 Preliminary Screening of ESUE Indicators

ESUE mainly measures the quality of ecological space and covers various ecological services provided by ecological space. In this study, the preliminary selection perspective of measurement indicators for ESUE mainly considers its inherent resources, human-intervened green infrastructure, human-land interaction, and

[5] The detailed meanings of the indicator abbreviations in the figure are listed in Table 6.5.

ecological services supplied by ecological space. In addition, consistent with the preliminary selection logic of measurement indicators for PSUE, the perspectives of landscape patches and pollutants are also taken into account. Accordingly, the perspectives for preliminary input indicators include resources, green infrastructure resources, and population; the perspectives for preliminary output indicators cover human-land relationship, ecological services, landscape, and pollutants. As shown in Table 6.6, a total of 28 indicators are initially selected for measuring ESUE, including 13 input indicators and 15 output indicators.

The data sources and calculation methods of indicators have been specified in Table 6.6. For indicators with straightforward calculation methods and those elaborated in the previous sections, no redundant elaboration will be provided in this study. The calculation methods of partial indicators are presented as follows.

1. **Calculation of Biological Abundance Index**

In accordance with the Technical Specification for Eco-environmental Status Evaluation (Trial) H/T192-2009 issued by the former State Environmental Protection Agency of China, the formulas are defined as Eqs. (6.5)–(6.6).

$$\begin{aligned}&\text{Biological Abundance Index}\\&= \text{A\_bio} \times (0.35 \times \text{Forest Area} + 0.21\\&\times \text{Grassland Area} + 0.28 \times \text{Water and Wetland Area}\\&+ 0.11 \times \text{Cultivated Area} + 0.04 \times \text{Construction Land Area} + 0.01\\&\times \text{Unused Land Area})/\text{Regional Area}\end{aligned} \tag{6.5}$$

$$\text{A}_{\text{bio}} = 100/\text{A}_{\max} \tag{6.6}$$

where, A_{bio} represents the normalization coefficient of the biological abundance index, and A_{max} is the maximum value of the biological abundance index before normalization. The reference value of A_{bio} is 511.26. In addition, all mentioned area units are square kilometers.

2. **Calculation of Ecosystem Service Value**

The calculation of ecosystem service value adopts the improved equivalent factor method per unit area proposed by Xie et al. [8]. The formulas are presented as Eqs. (6.7)–(6.8).

$$\text{ESV} = \sum_{i=1}^{n} \text{A}_i \times \text{VC}_i \tag{6.7}$$

$$\text{VC}_i = \sum_{i=1}^{n} \text{EC}_j \times \text{E}_a \tag{6.8}$$

Table 6.6 Preliminary indicators for measuring ecological space use efficiency

Indicator type	Indicator perspective	Specific indicator	Abbreviation	Indicator description	Data source and calculation method
Input Indicators	Resource Input (Natural and Non-natural Resources)	Mean DEM (m)	In_DEM	Mean value of DEM within the DMU	Calculated independently based on DEM grid data
		Ecological Space Score	In_StVa	Evaluation score of ecological space within the DMU	Calculated independently (refer to Sect. 6.2.1.1)
		Urban Green Area (ha)	In_GrAr	Total urban green area within the DMU	Statistical data
		Total Energy Consumption (ton)	In_GenCo	Total energy consumption within the DMU	Calculated independently from statistical data
		Green Coverage Rate of Built-up Area (%)	In_GcoRa	Ratio of total urban green area to built-up area within the DMU	Calculated independently from statistical data
	Green Infrastructure Resource Input	Core Area (m^2)	In_GcoAr	Core area within the DMU	Calculated independently based on land use data
		Bridge Area (m^2)	In_GbriAr	Bridge area within the DMU	Calculated independently based on land use data
		Islet Area (m^2)	In_GisAr	Islet area within the DMU	Calculated independently based on land use data
		Edge Area (m^2)	In_GedAr	Edge area within the DMU	Calculated independently based on land use data
		Perforation Area (m^2)	In_GpeAr	Perforation area within the DMU	Calculated independently based on land use data

(continued)

Table 6.6 (continued)

Indicator type	Indicator perspective	Specific indicator	Abbreviation	Indicator description	Data source and calculation method
		Loop Area (m^2)	In_GloAr	Loop area within the DMU	Calculated independently based on land use data
		Branch Area (m^2)	In_GbraAr	Branch area within the DMU	Calculated independently based on land use data
	Population Input	Total Population (10^4 persons)	In_Popu	Total population of the DMU during the research period	Statistical data
Output Indicators	Human-Land Output	Per Capita Core Area (m^2/person)	Out_GcoAar	Ratio of core area to total population within the DMU	Calculated independently based on land use data and statistical data
		Per Capita Green Space Area (m^2/person)	Out_GrAar	Ratio of urban green space area to total population within the DMU	Calculated independently from statistical data
	Ecological Service Output	Biological Abundance Index	Out_BiRi	Total biological abundance within the DMU	Calculated independently based on land use data (see below)
		Ecosystem Service Value (CNY)	Out_EseVa	Total ecosystem service value within the DMU	Calculated independently based on land use data (see below)
		Carbon Sequestration Capacity (ton)	Out_CaSi	Total carbon sequestration within the DMU	Calculated independently based on land use data (refer to Sect. 6.2.1.1)
		Soil Moisture Index	Out_WET	Average soil moisture of all grids within the DMU	Calculated independently on the Google Earth Engine platform

(continued)

Table 6.6 (continued)

Indicator type	Indicator perspective	Specific indicator	Abbreviation	Indicator description	Data source and calculation method
		Greenness Index	Out_NDVI	Average greenness of all grids within the DMU	Calculated independently on the Google Earth Engine platform
		Heat Index	Out_LST	Average heat intensity of all grids within the DMU	Calculated independently on the Google Earth Engine platform
		Dryness Index	Out_NDBI	Average dryness of all grids within the DMU	Calculated independently on the Google Earth Engine platform
	Landscape Perspective Output	Landscape Division Index	Out_DIVI	Division index within the DMU	Calculated independently based on land use data
		Effective Mesh Size	Out_MESH	Effective mesh size within the DMU	Calculated independently based on land use data
		Splitting Index	Out_SPLI	Splitting index within the DMU	Calculated independently based on land use data
		Shannon Diversity Index	Out_SHDI	Shannon diversity index within the DMU	Calculated independently based on land use data
		Shannon Evenness Index	Out_SHEI	Shannon evenness index within the DMU	Calculated independently based on land use data
	Pollutant Output	Total $PM_{2.5}$ Mass	Out_$PM_{2.5}$	Total $PM_{2.5}$ concentration within the DMU	Processed independently based on product data (see below)

where, ESV denotes ecosystem service value (CNY/a). *i* represents land use type. *j* represents ecosystem service type. A_i is the area of the *i*-th land use type (hm^2). VC_i is the ecosystem service value per unit area of the *i*-th land use type (CNY/($hm^2 \cdot a$)). EC_j refers to the ecosystem service value equivalent of a certain land use type. E_a is the economic value of one unit of ecosystem service equivalent (CNY/($hm^2 \cdot a$)). Considering the temporal characteristics of the research period and referring to existing academic studies, this study uniformly adopts the 2007 ecosystem service value equivalent table for statistical analysis, as shown in Table 6.7. Meanwhile, the economic value per unit ecosystem service equivalent is calculated at 3406.50 CNY/hm^2 based on the standard of 2010. The corresponding ecosystem service value per unit area across China is presented in Table 6.8.

3. **Calculation of $PM_{2.5}$**

The $PM_{2.5}$ data adopted in this study are derived from the $PM_{2.5}$ dataset covering 1960–2020 published by Zhong et al., which is further processed through clipping, statistical analysis, calculation, and other corresponding procedures [9].

Table 6.7 Ecosystem service value equivalent per unit area (2007)

		Forestland	Grassland	Cultivated land	Water area	Unused land
Provisioning services	Food production	0.93	1.23	1	0.77	0
	Raw material production	2.73	0.19	0.1	0.11	0.01
Regulating services	Gas regulation	0.65	0.29	0.13	0.06	0.17
	Climate regulation	3.1	1.14	1	0.68	0.12
	Hydrological regulation	2.36	0.19	0.13	122.98	0
	Waste treatment	1.78	1.6	0.17	0.1	0.01
Supporting services	Soil conservation	8.86	2.41	1.7	0.22	0
	Biodiversity maintenance	3.8	3.6	1	0.87	0.07
Cultural services	Aesthetic landscape provision	1.8	0.36	0.06	3.97	0.02
Total		26.01	11.01	5.29	129.76	0.4

Table 6.8 Ecosystem service value per unit area of Chinese ecosystems (CNY/(hm^2·a)) (2010)

Ecosystem service value		Forestland	Grassland	Cultivated land	Water area	Unused land
Provisioning services	Food production	3168.045	4189.995	3406.5	2623.005	0
	Raw material production	9299.745	647.235	340.65	374.715	34.065
	Subtotal	12,467.79	4837.23	3747.15	2997.72	34.065
Regulating services	Gas regulation	2214.225	987.885	442.845	204.39	579.105
	Climate regulation	10,560.15	3883.41	3406.5	2316.42	408.78
	Hydrological regulation	8039.34	647.235	442.845	418,931.37	0
	Waste treatment	6063.57	5450.4	579.105	340.65	34.065
	Subtotal	26,877.29	10,968.93	4871.295	421,792.83	1021.95
Supporting services	Soil conservation	30,181.59	8209.665	5791.05	749.43	0
	Biodiversity maintenance	12,944.7	12,263.4	3406.5	2963.655	238.455
	Subtotal	43,126.29	20,473.07	9197.55	3713.085	238.455
Cultural services	Aesthetic landscape provision	6131.7	1226.34	204.39	13,523.805	68.13
Total		88,603.065	37,505.565	18,020.385	442,027.44	1362.6

6.2.3.2 ESUE Indicator Calibration via GeoFA Model

1. **Input Indicators**

The GeoFA model is adopted to analyze thirteen initially selected input indicators for ESUE (see Table 6.6), and the common factor loading matrix was obtained, as shown in Fig. 6.7. From the resource perspective, the green coverage rate of built-up area (In_GcoRa), total energy consumption (In_GenCo), urban green area (In_GrAr), and mean DEM (In_DEM) exhibit high loading values in the common factors. Among them, In_GcoRa ranks highest in corresponding factor loadings of Common Factor 4 in each year, with markedly higher loadings than other indicators within this common factor. The seven indicators from the green infrastructure resource perspective all present high loading values in the common factors. Branch area (In_GbraAr) and perforation area (In_GpeAr) rank first in the loading values of multiple common factors, including Common Factor 1 and Common Factor 3 in 1990, 2000, 2010, and 2020. From the population perspective, total population (In_Popu) shows relatively high loading values and ranks first in Common Factor 2 of 2000 and

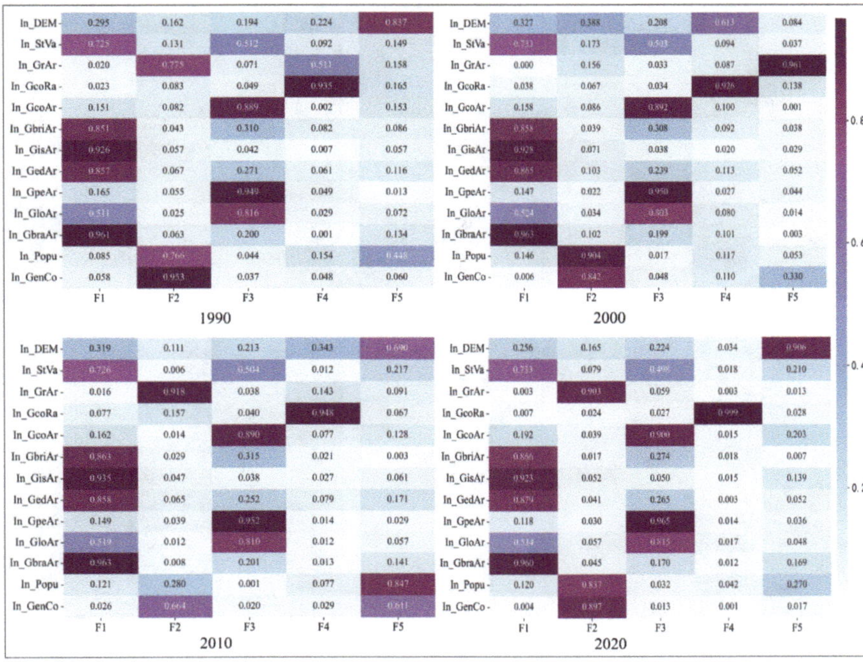

Fig. 6.7 Heatmap of factor loading matrix for input indicators of ecological space use efficiency[6]

Common Factor 5 of 2010. Based on the comprehensive consideration, In_GcoRa, In_GenCo, In_GbraAr, In_GpeAr, and In_Popu are finally selected as the input indicators for ESUE.

2. **Output Indicators**

This study adopts the GeoFA model to analyze fifteen initially selected output indicators of ESUE (see Table 6.6) and obtained the common factor loading matrix, as illustrated in Fig. 6.8. From the human-land perspective, per capita green space area (Out_GrAar) and per capita core area (Out_GcoAar) both exhibit high loading values in common factors. Specifically, per capita green space area (Out_GrAar) ranks first in the loading values of Common Factor 5 in 1990 and 2010, as well as Common Factor 4 in 2000 and 2020. From the ecosystem service perspective, ecosystem service value (Out_EseVa), dryness index (Out_NDBI), and land surface temperature index (Out_LST) rank first in the loading values of corresponding common factors, demonstrating the essential role of ecosystem service indicators in characterizing ecological space development. From the landscape perspective, Shannon Evenness Index (Out_SHEI) and Splitting Index (Out_SPLI) exhibit prominent factor loadings in common factors. Out_SHEI occupies the top loading value of Common Factor 2 in 1990, 2000, and 2010, while Out_SPLI ranks first in Common Factor 6 in 1990,

[6] The specific connotations of the indicator abbreviations in the figure are shown in Table 6.6.

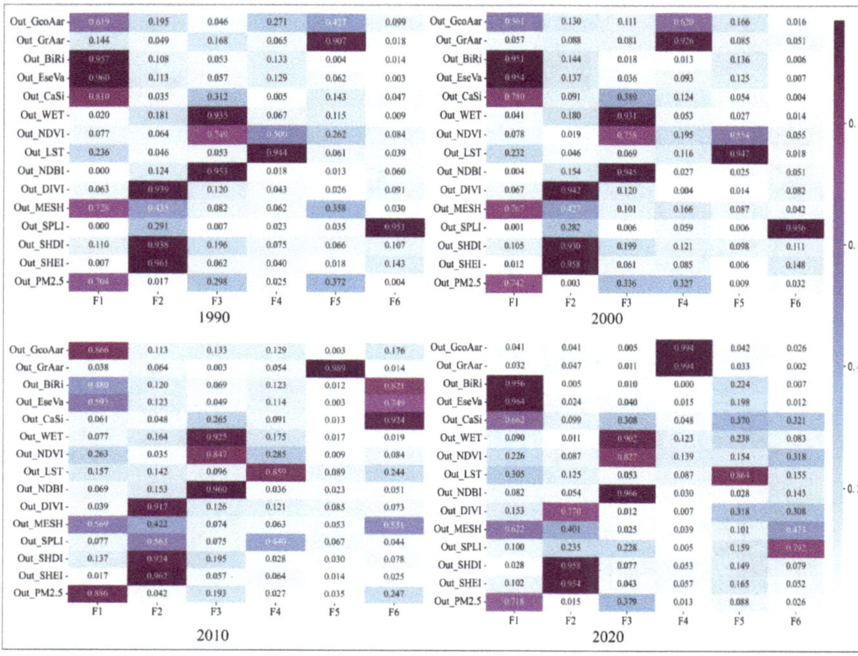

Fig. 6.8 Heatmap of factor loading matrix for output indicators of ecological space use efficiency[7]

2000, and 2020. After comprehensive consideration, Out_GrAar, Out_EseVa, Out_SHEI, Out_NDBI, Out_LST and Out_SPLI are ultimately selected as the output indicators for ESUE. The first three indicators belong to desirable outputs, and the latter three are undesirable outputs.

In conclusion, this study adopts the perspective of production, living, and ecological space use. It applies the GeoFA model to perform dimensionality reduction and screening analysis on dozens of preliminary indicators. This paper finally establishes an indicator system of TSUE at the municipal scale across China. The research provides a solid foundation for subsequent efficiency measurement and prediction analysis.

6.2.4 Dimensional Integration for Comprehensive TSUE Calculation

This study measures the use efficiency of production space, living space, and ecological space. On this basis, it further determines the overall TSUE. The calculation formula is presented as Eq. (6.9).

[7] The detailed definitions of the indicator abbreviations in the figure are listed in Table 6.6.

$$\mathrm{E}_{\mathrm{all}} = \sum_{i}^{n} \mathrm{E}_i \mathrm{W}_i \tag{6.9}$$

With the formula, E_{all} refers to the overall TSUE. E_i denotes the PSUE, LSUE, and ESUE, respectively. W_i represents the corresponding weight value. The entropy weight method is adopted in this study to calculate weights for the three types of spatial use efficiency. The weight values of production space, living space, and ecological space are 0.4431, 0.4272, and 0.1297 correspondingly. This study calculates the overall TSUE for 1990, 2000, 2010, and 2020. The calculation relies on the use efficiency values and weight coefficients of production, living, and ecological space in the four periods.

6.3 Spatiotemporal Evolution and Spatial Autocorrelation Analysis of Territorial Space Use Efficiency

6.3.1 Spatiotemporal Evolution of TSUE

This study adopts the GeoDEA model constructed in Chapter 4 to measure the TSUE of 340 cities across China in 1990, 2000, 2010, and 2020. This section focuses on the measurement of historical and current efficiency values and the analysis of their spatiotemporal evolution characteristics. The large number of research cities hinders inter-city efficiency comparison of efficiency values among different cities and clear visual expression. The study classifies TSUE values into five grades in ascending order. The five grades include low-efficiency level below 0.36, medium–low efficiency level from 0.36 to 0.72, medium-efficiency level from 0.72 to 1.08, medium–high efficiency level from 1.08 to 1.44, and high-efficiency level above or equal to 1.44. Relevant analyses are carried out on the basis of the above classification standards.

6.3.1.1 Spatiotemporal Evolution of PSUE (1990–2020)

From the temporal perspective, Table 6.9 presents the statistical quantity of cities at different levels of PSUE during 1990–2020. Three major characteristics are summarized through data analysis.

1. The number of cities at various efficiency levels shows obvious fluctuations in each year. The overall efficiency level follows a changing trend of rise, decline, and rebound. From 1990 to 2000, the overall efficiency level shows an upward trend. The number of cities at the low-efficiency level drops rapidly from 71 to 37. The quantity of cities at the medium–low efficiency level rises remarkably from 46 to 97, with its proportion increasing by 15 percentage points. The number

Table 6.9 Statistics on the number of cities at different production space use efficiency levels, 1990–2020 (unit: number)

Year	Low-efficiency level	Medium–low efficiency level	Medium-efficiency level	Medium–high efficiency level	High-efficiency level
1990	71(20.88%)	46(13.53%)	111(32.65%)	82(24.12%)	30(8.82%)
2000	37(10.88%)	97(28.53%)	110(32.35%)	77(22.65%)	19(5.59%)
2010	148(43.53%)	6(1.76%)	82(24.12%)	84(24.71%)	20(5.88%)
2020	36(10.59%)	74(21.76%)	146(42.94%)	71(20.88%)	13(3.82%)

of cities at medium, medium–high, and high-efficiency levels declines slightly. A distinct downward trend appears in the overall efficiency level from 2000 to 2010. Obvious quantity changes occur among cities at low, medium–low, and medium-efficiency levels. Compared with the data in 2000, the number of low-efficiency cities surges by 111, with the proportion rising by approximately 32.65 percentage points. The quantity of cities at medium–low and medium-efficiency levels decreases sharply. Cities at medium–high and high-efficiency levels maintain a stable scale. PSUE improves to a certain extent from 2010 to 2020. The number of cities at the low-efficiency level decreases significantly. The quantity of cities at medium–low and medium-efficiency levels increases by 68 and 64, respectively. By contrast, the number of cities at medium–high and high-efficiency levels presents a downward trend.

2. The year 2010 constitutes a critical turning point of long-term changes in PSUE. This phenomenon is closely related to the staged characteristics of China's economic development. Since the beginning of the twenty-first century, the national economy has followed a development path driven by capital factors and entered a period of rapid economic growth accompanied by profound transformation of economic structure. The level of scientific and technological development remains relatively backward in this period. Excessive resource input and resource waste inevitably emerged and further reduced PSUE. After 2010, China's economic growth began to slow down. The economic development entered a new stage featuring growth rate shift, economic restructuring, and resolution of accumulated developmental problems. This stage emphasizes development quality with a reasonable input–output ratio. Corresponding improvement can be observed in PSUE.
3. The number of cities at low, medium–low, and medium-efficiency levels presents dramatic fluctuations over the long-time series. The number of cities at medium–high and high-efficiency levels fluctuates mildly yet shows an overall declining tendency. Cities with low-efficiency levels are characterized by immature development and poor stability. Such cities are susceptible to policy adjustments, structural optimization, resource allocation, and technological innovation. Changes in the overall socio-economic environment can trigger substantial fluctuations in their PSUE. Cities with high-efficiency levels maintain sound development

conditions with mature industrial structure and advanced scientific and technological innovation capacity. These cities possess strong overall stability and anti-risk capability, which keeps their quantity relatively steady. Changes in the overall socio-economic environment still exert subtle influences and bring slight variations to their evolutionary trend.

From the spatial perspective, this study conducts visual graded presentation of PSUE in each year to fully reveal the characteristics of spatiotemporal patterns. As shown in Fig. 6.9, the spatial development of PSUE presents unbalanced distribution and forms a complex spatial pattern. The overall efficiency level remains relatively low in northern, western, and southern regions, while it stays relatively high in central and eastern regions. Most cities in Xinjiang Uygur Autonomous Region and Inner Mongolia Autonomous Region belong to the low-efficiency level. Restricted by natural background conditions, climatic environment, historical and cultural factors, as well as innovation capacity, local economic industries are dominated by the primary and secondary industries. These regions develop at a slower pace compared with other areas in China. The unbalanced input–output ratio further leads to low-efficiency values. Some cities in southern China also maintain a low level of PSUE. Liuzhou and Guilin in Guangxi serve as typical examples. The pillar industries of the two cities face prominent drawbacks including relatively backward industrial foundation, weak technological innovation capacity, underdeveloped high-tech industries, and insufficient market competitiveness. These factors greatly affect the input–output relationship and result in low PSUE. Central China and eastern coastal areas represent regions with high economic development levels in China. Such regions possess rational industrial structure, strong scientific and technological innovation capability, and favorable opening-up conditions. The combined effect of these advantages optimizes the development environment of production space and contributes to high use efficiency.

6.3.1.2 Spatiotemporal Evolution of LSUE (1990–2020)

From the temporal perspective, Table 6.10 shows the statistics of the number of cities at different levels of LSUE during 1990–2020. Three typical characteristics are summarized through analysis.

1. The number of cities at different efficiency levels fluctuates annually, and the overall efficiency level follows an evolutionary trend of decline, rise, and further decline. From 1990 to 2000, the overall efficiency level presents a downward trend. The number of cities at the medium-efficiency level decreases from 108 in 1990 to 76 in 2000, with the proportion dropping by 9.41 percentage points. The number of cities at other efficiency levels changes within a small range. From 2000 to 2010, the overall efficiency level shows a notable upward trend. The number of cities at the low-efficiency level drops sharply from 115 in 2000 to 72 in 2010, with the proportion falling by 12.64 percentage points. The number of cities at the medium–low efficiency level decreases from 71 to 61. The number of

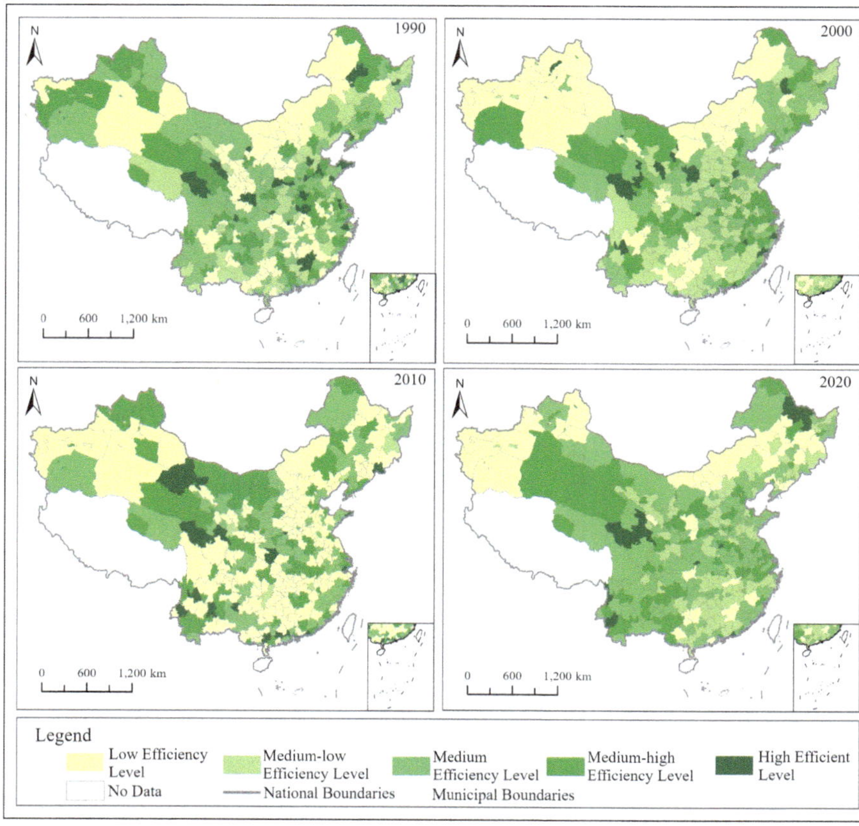

Fig. 6.9 Spatial distribution of production space use efficiency from 1990 to 2020[8]

Table 6.10 Statistics of cities at different living space use efficiency levels, 1990–2020 (unit: number)

Year	Low-efficiency level	Medium–low efficiency level	Medium-efficiency level	Medium–high efficiency level	High-efficiency level
1990	114(33.53%)	51(15.00%)	108(31.76%)	60(17.65%)	7(2.06%)
2000	115(33.82%)	71(20.88%)	76(22.35%)	70(20.59%)	8(2.35%)
2010	72(21.18%)	61(17.94%)	132(38.82%)	66(19.41%)	9(2.65%)
2020	96(28.24%)	68(20.00%)	95(27.94%)	74(21.76%)	7(2.06%)

cities at the medium-efficiency level increases rapidly from 76 to 132, accounting for 38.82% of all cities in 2010. From 2010 to 2020, the overall efficiency level shows a downward tendency. This trend is reflected in a marked increase of low-efficiency cities and a prominent decrease of medium-efficiency cities. The

[8] This figure is compiled based on the base map with the review approval number GS (2019) 1822.

number of low-efficiency cities rises from 72 to 96, with the proportion growing by 7.06 percentage points. The number of medium-efficiency cities declines from 132 to 95, with the proportion decreasing by 10.88 percentage points.

2. The year 2010 acts as the turning point for the long-term variation of LSUE. China's economy maintains rapid development from 2000 to 2010. Economic growth, abundant employment opportunities, continuous improvement of infrastructure and stable living conditions jointly promote the gradual improvement of LSUE. Economic growth slows down after 2010. Industrial optimization, resource integration and policy adjustment exert certain influences on the layout of living space. The spread of epidemic viruses also affects residents' daily life to a certain extent. These factors jointly lead to a downward trend of LSUE during 2010–2020.
3. The number of cities at low, medium–low, and medium-efficiency levels experiences dramatic fluctuations over the long-time series. The variation range remains small for cities at medium–high and high-efficiency levels. Cities with low-efficiency levels lack mature and complete living supporting facilities. Such conditions reduce the convenience of living space and render medical and educational services insufficient to meet social demands. These deficiencies further trigger population outflow and eventually result in drastic fluctuations in LSUE. In contrast, cities with high LSUE are equipped with comprehensive living service facilities. These cities well satisfy public demands in education, medical care, leisure, and entertainment. A stable population scale coordinates the input and output of living space. Such a balanced development state keeps the efficiency fluctuation within a small range.

From the spatial perspective, this study presents visualized hierarchical results of LSUE for each year to fully reveal the characteristics of spatiotemporal patterns. As shown in Fig. 6.10, the spatial distribution of LSUE shows obvious imbalance. Parts of cities in Northeast China, Central China, and Southwest China maintain relatively low-efficiency levels. Low-efficiency cities in Central China and Southwest China exhibit prominent spatial agglomeration, while Northwest China and Northern China generally present higher efficiency levels. In Northeast China, low levels of LSUE appear in the Liaodong Peninsula in 2000, most parts of Northeast China in 2010, and southern Heilongjiang in 2020. Economic recession in Northeast China has slowed regional economic growth. The trend undermines employment opportunities and resident income, triggers population loss, and further restricts the improvement of LSUE. Low-efficiency cities in Central China show obvious spatial agglomeration. The agglomeration area was prominent and covered a large number of cities in 2000, mainly concentrated in Hubei, Hunan, and Henan. Central cities achieve rapid economic expansion and substantial population growth with active population mobility compared with other regions. The construction and coordinated development of living infrastructure fail to keep pace with growing public demands, which results in low LSUE. Southwest China forms an agglomeration of low-efficiency cities mainly concentrated in Yunnan Province. Border location and harsh natural conditions lead to weak economic foundation, underdeveloped transportation, and

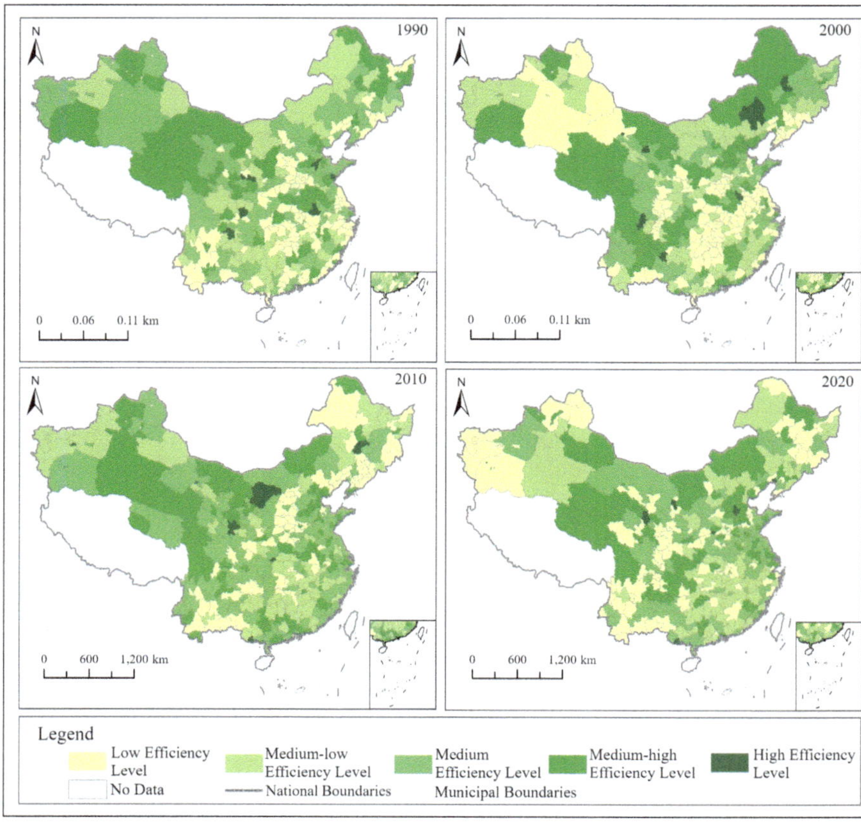

Fig. 6.10 Spatial distribution of living space use efficiency from 1990 to 2020[9]

inadequate educational resources in Yunnan. Such problems contribute to a high incidence and severe degree of poverty across the region. Official statistics on national poverty counties show that Yunnan contains 88 national-level poverty-stricken counties, among which 27 are classified as extreme poverty counties. Such underdevelopment keeps local LSUE at a low level. Western China and Northern China feature vast territory, abundant natural resources, small population size, and low population density. The region enjoys sufficient per capita resources and stable living conditions for residents. Such advantages support the maintenance of a relatively high level of LSUE.

[9] This figure is compiled based on the base map with the review approval number GS (2019) 1822.

6.3.1.3 Spatiotemporal Evolution of ESUE (1990–2020)

From the temporal dimension, Table 6.11 presents the statistical number of cities at different levels of ESUE from 1990 to 2020. Three features can be identified from statistical results.

1. The number of cities at different efficiency levels shows slight fluctuations in each year. The overall efficiency level presents a steady upward trend. From 1990 to 2000, the number of cities at the low-efficiency level remains stable. Cities at the medium–low efficiency level experience minor changes. The number of cities at the medium-efficiency level increases obviously from 134 to 165, accounting for 48.53% of the total cities in 2000. The number of cities at medium–high and high-efficiency levels both declines slightly. From 2000 to 2010, the quantity of cities at all efficiency levels changes slightly and maintains a steady growth trend. The number of cities at the low-efficiency level decreases by 9, and the number of cities at the medium–low efficiency level increases by 9. The number of cities at the medium-efficiency level rises by 8, reaching a proportion of 50.88% in 2010. The quantity of cities at the medium–high efficiency level remains unchanged. From 2010 to 2020, the number of cities at the low-efficiency level drops to 4, accounting for only 1.18% of the total sample. The number of cities at the medium–low efficiency level decreases sharply from 66 in 2010 to 31 in 2020. The number of cities at the medium-efficiency level maintains continuous growth from 173 to 196, with its proportion rising to 57.65% in 2020. The number of cities at the medium–high efficiency level increases by 17, with its proportion growing by 5 percentage points. The quantity of cities at the high-efficiency level keeps constant.
2. The number of cities at the high-efficiency level shows a declining trend. ESUE maintains an overall steady growth tendency. In-depth analysis of the datasets from 1990 to 2020 reveals a remarkable drop in the number of high-efficiency cities by 50%, falling from 30 in 1990 to 15 in 2020. The result reflects prominent difficulties in the efficient utilization of ecological space across China. More in-depth research is required on the coordination relationship between human and nature as well as reasonable development boundaries. Green economy should be developed under the premise of sound ecological environment and long-term

Table 6.11 Statistics on the number of cities at different ecological space use efficiency levels, 1990–2020 (unit: number)

Year	Low-efficiency level	Medium–low efficiency level	Medium-efficiency level	Medium–high efficiency level	High-efficiency level
1990	16(4.71%)	67(19.71%)	134(39.41%)	93(27.35%)	30(8.82%)
2000	18(5.29%)	57(16.76%)	165(48.53%)	77(22.65%)	23(6.76%)
2010	9(2.65%)	66(19.41%)	173(50.88%)	77(22.65%)	15(4.41%)
2020	4(1.18%)	31(9.12%)	196(57.65%)	94(27.65%)	15(4.41%)

sustainability. Ecological space needs rational utilization without exceeding the maximum threshold of ecological carrying capacity.

3. Cities at the medium-efficiency level occupy a dominant proportion. The share of cities at the medium level of ESUE remains relatively high and keeps rising from 1990 to 2020. The proportion is 39.41% in 1990 and increases to 57.65% in 2020, with a total growth of 18.24 percentage points.

From the spatial perspective, this study conducts visual hierarchical display of ESUE for each year to fully reveal the characteristics of spatiotemporal patterns, as illustrated in Fig. 6.11. The spatial pattern of ESUE presents an obvious distribution feature of high values in the west and south, and low values in the east and north. Specifically, cities in Xinjiang, Qinghai, northwestern Gansu, western Sichuan and western Yunnan maintain relatively high levels of ESUE. These areas overlap with China's First Topographic Ladder. They feature high altitude and superior original ecological conditions and serve as important water source conservation areas in China. Meanwhile, economic development in these regions progresses at a relatively slow pace. Human disturbance and damage to the ecological environment remain limited, which sustains a high level of ESUE. Eastern China has low terrain and rapid economic growth. Industrial and domestic pollution impose certain impacts on ecological conditions, leading to lower ESUE compared with western regions. Northern China including Inner Mongolia, northern Hebei and western Liaoning suffers harsh climatic conditions. Sandstorms from Mongolia and prevailing northerly winds restrain local ecological conditions and result in low ESUE. Southern China enjoys humid climate, moderate temperature, and relatively high forest coverage. The region also maintains sound economic development and rational urban planning. Its ecological network composed of green infrastructure is well constructed and coordinated. These advantages jointly keep ESUE at a relatively high level.

6.3.1.4 Spatiotemporal Evolution of Overall TSUE (1990–2020)

From the temporal perspective, Table 6.12 presents the statistics on the number of cities at different levels of overall TSUE during 1990–2020. Four core characteristics are summarized based on analytical results.

1. The number of cities at various efficiency levels fluctuates across years. The overall efficiency level follows an evolutionary trend of growth, decline, and renewed growth. From 1990 to 2000, the overall efficiency level shows an upward trend. The number of low-efficiency cities decreases by 11, with its proportion falling by 3.24 percentage points. The number of medium–low and medium-efficiency cities increases by 16 and 13, respectively. Medium–high efficiency cities decline from 67 to 51, with the proportion dropping by 4.71 percentage points. High-efficiency cities only experience slight numerical change. From 2000 to 2010, the overall efficiency level declines markedly. The number of low-efficiency cities surges from 18 to 53, with the proportion rising by 10.3

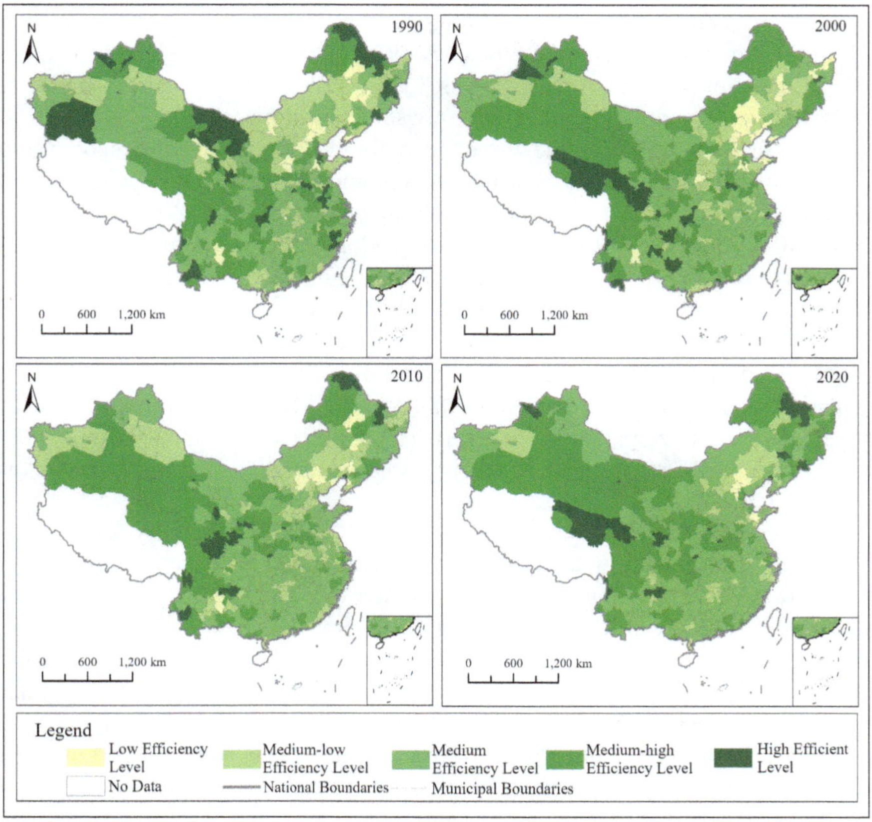

Fig. 6.11 Spatial distribution of ecological space use efficiency from 1990 to 2020[10]

Table 6.12 Statistics on the number of cities at different levels of overall territorial space use efficiency, 1990–2020 (unit: number)

Year	Low-efficiency level	Medium–low efficiency level	Medium-efficiency level	Medium–high efficiency level	High-efficiency level
1990	29(8.53%)	113(33.24%)	121(35.59%)	67(19.71%)	10(2.94%)
2000	18(5.29%)	129(37.94%)	134(39.41%)	51(15.00%)	8(2.35%)
2010	53(15.59%)	111(32.65%)	107(31.47%)	62(18.24%)	7(2.06%)
2020	11(3.24%)	119(35.00%)	145(42.65%)	58(17.06%)	7(2.06%)

percentage points. Medium–low and medium-efficiency cities decrease sharply by 18 and 27, respectively. Medium–high efficiency cities increase from 51 to 62. From 2010 to 2020, the efficiency level shows a noticeable improvement. Low-efficiency cities drop sharply from 53 to 11, with the proportion falling by

[10] This figure is compiled based on the base map with the review approval number GS (2019) 1822.

12.35 percentage points and accounting for merely 3.24% in 2020. The number of medium–low and medium-efficiency cities increases by 8 and 38 separately, and medium-efficiency cities account for 42.65% in 2020. Medium–high and high-efficiency cities maintain relatively stable quantities.

2. The year 2010 constitutes the turning point for the long-term evolution of overall TSUE. In accordance with the national economic development path, China maintains rapid economic growth from 2000 to 2010 with an average annual GDP growth rate above 10%. The extensive development pattern during this period leads to low land intensification, excessive resource input, imperfect infrastructure, and natural environmental pollution. These problems result in a low level of overall TSUE. After 2010, economic development shifts from high-speed growth to a moderate growth stage. A series of problems accumulated during the earlier rapid expansion are addressed through policy adjustment, rational resource allocation, and technological progress. The optimization of input–output structure realizes the maximum utilization of various resources and further improves overall TSUE.
3. The number of cities at low, medium–low, and medium-efficiency levels experiences substantial fluctuations over the study period. Fluctuations remain moderate for cities at medium–high and high-efficiency levels. Cities with low overall TSUE are still in the adaptive development stage in production activities, living services, and human-environmental coordination. Such cities feature internal variability and instability. External disturbances such as policy guidance and economic changes further drive diverse evolutionary trends in their TSUE. By contrast, cities with high overall TSUE achieve sound coordination among production, living, and ecological subsystems. These subsystems reach a balanced, constrained, and coordinated development state, thereby maintaining relatively stable overall efficiency levels.
4. The temporal variation characteristics of overall TSUE are consistent with those of PSUE. This consistency indicates that production activities exert a dominant influence on overall territorial space use. Production activities generate substantial interference with land resources, energy resources, and ecological resources. Such interference indirectly affects the utilization process of living and ecological space and ultimately determines the overall level of TSUE.

From the spatial perspective, this study conducts visual graded presentation of overall TSUE for each year to fully characterize its spatiotemporal pattern, as shown in Fig. 6.12. The spatial development of overall TSUE is unbalanced and features a complex and volatile distribution pattern. In 1990, cities at the low-efficiency and medium–low efficiency levels were mainly distributed in parts of Inner Mongolia, Hubei, and Hunan, while those at the medium–high efficiency and high-efficiency levels were concentrated in Northeast China and central provinces north of the Huai River. In 2000, low-efficiency cities were predominantly located in Xinjiang, Hubei, and Hunan, whereas medium–high efficiency and high-efficiency cities were mainly found in Qinghai, Gansu, Sichuan, and Shandong. In 2010, cities with different levels of overall TSUE presented an interleaved distribution, with low-efficiency

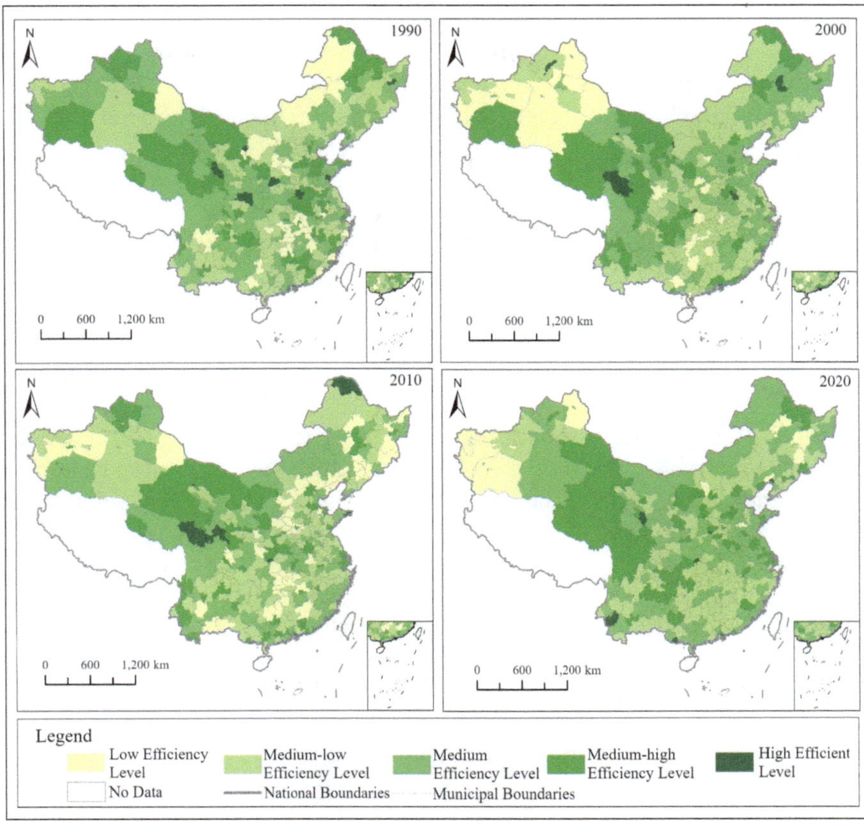

Fig. 6.12 Spatial distribution of overall territorial space use efficiency from 1990 to 2020[11]

and medium–low efficiency cities existing in all provinces. In 2020, low-efficiency cities were mainly concentrated in Xinjiang and Northeast China; medium–low efficiency cities were largely distributed in Inner Mongolia, Hubei, Hunan, Jiangxi, and other provinces; cities with relatively high-efficiency levels were primarily located in Qinghai, Hebei, Shandong, and Henan. Overall, eastern coastal cities maintain a high level of overall TSUE, while certain cities in northern, central, and western regions record relatively low overall efficiency. Furthermore, cities with low-efficiency and high-efficiency levels are spatially interlaced, and the efficiency level of individual cities has undergone evident changes over time. It is indicated that overall TSUE constitutes a complex research issue. The exploration of its evolutionary mechanism requires further in-depth investigation based on realistic conditions, such as defining a baseline year and establishing benchmark evaluation units.

Territorial space is a giant system involving complex human-land relationships, with its internal structures intertwined, mutually influenced, and mutually restricted.

[11] This figure is compiled based on the base map with the review approval number GS (2019) 1822.

Accordingly, the temporal changes and spatial patterns of TSUE exhibit complex characteristics. In summary, from the temporal perspective, the temporal evolution characteristics of PSUE, LSUE, and ESUE differ from one another. The year 2010 marks a critical turning point for the evolution of TSUE. The changing trend of overall TSUE is consistent with that of PSUE. From the spatial perspective, the spatial patterns of PSUE, LSUE, ESUE, and overall TSUE are complex and diversified. Overall, efficiency levels are generally low in the northwest, north, and southwest regions, while relatively high in central areas and coastal regions.

6.3.2 Spatial Autocorrelation of TSUE

According to the analytical results in Sect. 6.3.1, the PSUE, LSUE, ESUE, and overall TSUE of 340 cities across China from 1990 to 2020 all showed varying degrees of agglomeration characteristics in their spatial distribution patterns. In addition, given the geographical attributes of territorial space, certain spatial correlation and clustering features may emerge in the efficiency levels of neighboring cities. Therefore, to further quantitatively explore such characteristics and evolutionary rules, this section conducts global and local spatial autocorrelation analysis for the 340 prefecture-level cities nationwide on the basis of the findings in Sect. 6.3.1.

6.3.2.1 Global Spatial Autocorrelation of TSUE

First, this study adopts ArcGIS Pro 3.0 and GeoDa 1.22 to calculate and characterize the Global Moran's I index and corresponding scatter plots of PSUE, LSUE, ESUE, and overall TSUE across 340 cities in China. The spatial correlation of TSUE among these cities is analyzed from an overall perspective. The measurement results of the Global Moran's I index include three core indicators. The first is the Global Moran's I value, which ranges from -1 to 1. A value greater than 0 indicates positive spatial correlation of efficiency; a value less than 0 indicates negative spatial correlation; a value equal to 0 implies no spatial correlation. The second is the Z-score. A Z-score above 1.65 suggests spatially clustered distribution; a Z-score below -1.65 indicates spatially dispersed distribution; a Z-score between -1.65 and 1.65 means no significant spatial autocorrelation. The third is the *P*-value. A *P*-value less than 0.1 rejects the null hypothesis at the 90% confidence level, and a *P*-value less than 0.01 rejects the null hypothesis at the 99% confidence level, both verifying the existence of spatial autocorrelation. A *P*-value greater than 0.1 accepts the null hypothesis, indicating no spatial autocorrelation. In addition, the Global Moran's I scatter plot divides the entire spatial field into four quadrants corresponding to four types of spatial autocorrelation. The first quadrant represents High–High agglomeration, denoting high-efficiency cities surrounded by other high-efficiency cities. The second quadrant denotes Low–High outliers, where low-efficiency cities are surrounded by high-efficiency counterparts. The third quadrant refers to Low–Low agglomeration, indicating low-efficiency cities adjacent to other low-efficiency cities. The fourth quadrant represents High–Low outliers, where high-efficiency cities are surrounded by low-efficiency ones.

The Global Moran's I results show that the *P*-values of PSUE, LSUE, ESUE, and overall TSUE for the 340 cities from 1990 to 2020 are all lower than 0.01, passing the significance test at the 1% level. The null hypothesis of no spatial autocorrelation is firmly rejected for all four types of efficiency. Further analysis of Global Moran's I and Z-scores is presented in Figs. 6.13 and 6.14. All Global Moran's I values are positive and all Z-scores exceed 1.65, confirming significant positive spatial autocorrelation in TSUE across cities. As shown in Fig. 6.13, the Global Moran's I value of PSUE follows a trend of rise-decline-rise from 1990 to 2020, indicating that spatial correlation first strengthened, then weakened, and strengthened again, with 2010 as the turning point. The Global Moran's I of LSUE and overall TSUE both presented a trend of rise followed by decline, with the turning point occurring in 2000. The Global Moran's I value of ESUE experienced a trend of decline followed by rise with a mild growth range, revealing an overall weakening tendency in its spatial correlation.

This study further explores the agglomeration characteristics of TSUE across cities from the perspective of scatter plots. As shown in Fig. 6.15, the scatter plot of PSUE indicates that most data points fall within the High-High and Low-Low agglomeration quadrants during 1990–2020. This reflects that cities with high-efficiency values are spatially clustered with neighboring high-efficiency cities, and

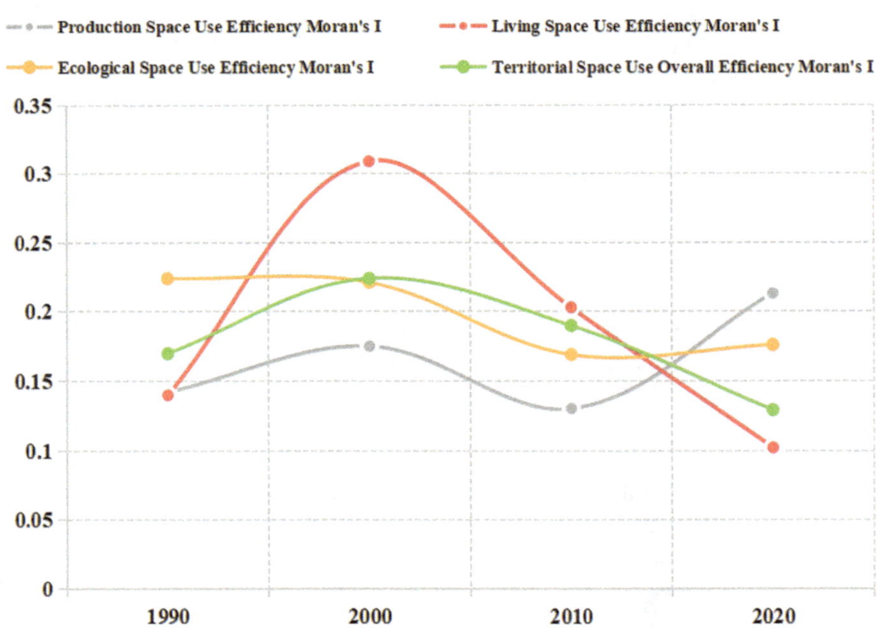

Fig. 6.13 Global Moran's I values of territorial space use efficiency from 1990 to 2020

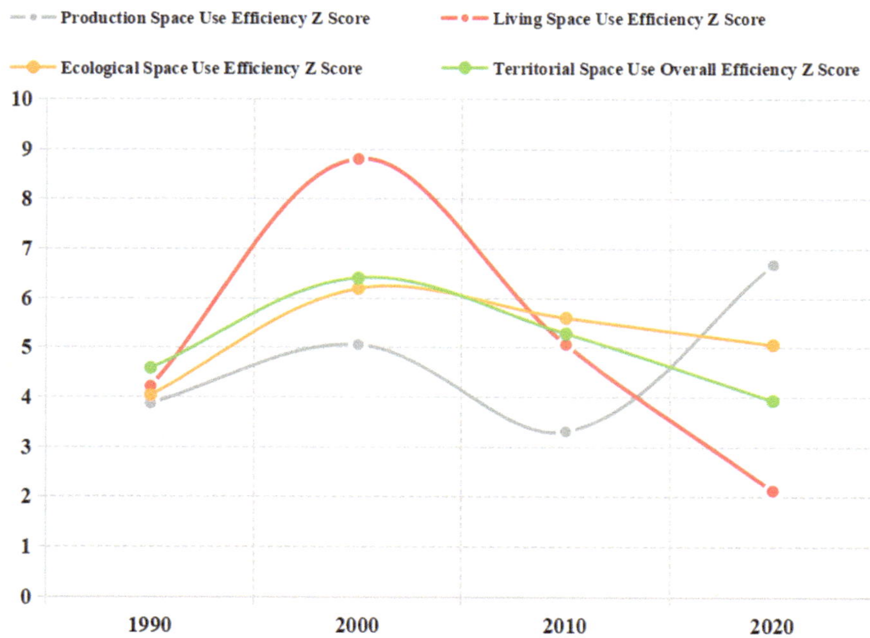

Fig. 6.14 Z-Scores of territorial space use efficiency from 1990 to 2020

cities with low-efficiency values are clustered with neighboring low-efficiency cities. Relatively few cities are distributed in the Low–High and High–Low outlier quadrants. This implies that the PSUE of individual cities does not differ substantially from that of their neighbors, further verifying the existence of spatial autocorrelation in PSUE.

By the same analytical logic, the scatter plot of LSUE is shown in Fig. 6.16. From 1990 to 2020, most data points are distributed in the High–High and Low–Low agglomeration quadrants, while only a small number of cities fall into the Low–High and High–Low outlier quadrants. This indicates that the LSUE of individual cities has no significant difference from that of their neighboring cities, further confirming the presence of spatial autocorrelation in LSUE. From the evolutionary changes of the scatter plots, the spatial agglomeration of LSUE gradually shifts from clustered to dispersed, especially during 2010–2020. In 2010, the spatial agglomeration of LSUE among the 340 cities was prominent, whereas outliers gradually increase by 2020.

Following the same analysis, the scatter plot of ESUE is presented in Fig. 6.17. From 1990 to 2020, most data points are concentrated in the High–High and Low–Low agglomeration quadrants, whereas fewer points fall in the Low–High and High–Low outlier quadrants with a gradually decreasing trend. This suggests that the ESUE of individual cities shows no obvious discrepancy from that of their neighboring cities, which further verifies significant spatial autocorrelation in ESUE. From the evolutionary changes of the scatter plots, outliers in the spatial agglomeration of

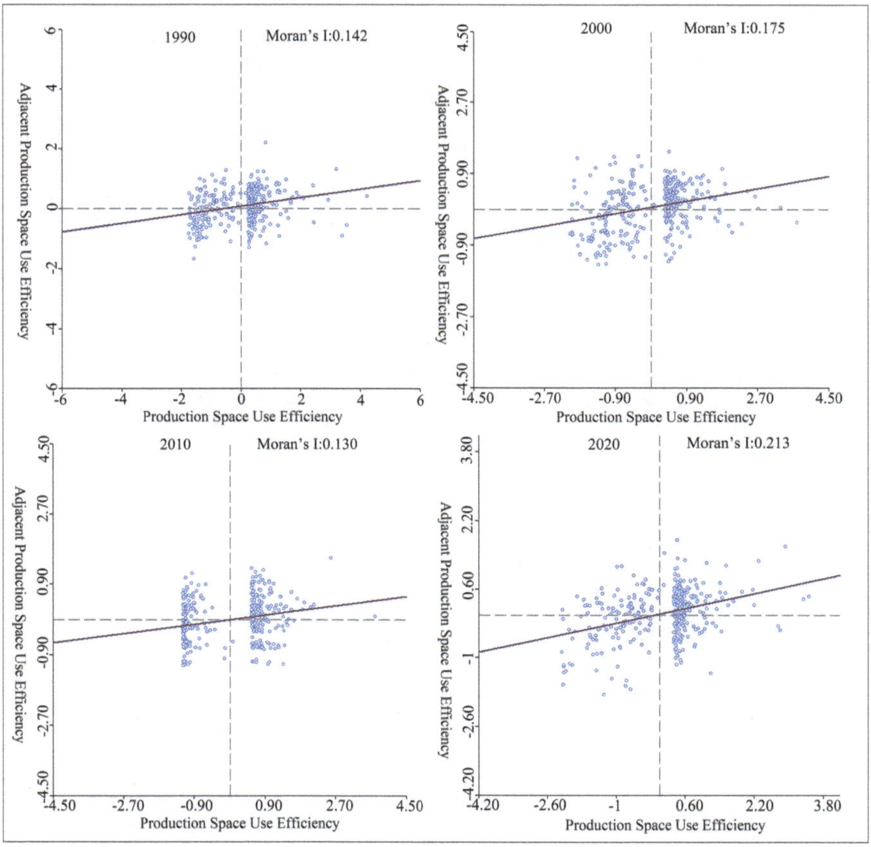

Fig. 6.15 Global Moran's I scatter plot of production space use efficiency

ESUE increased gradually during 1990–2020. Overall, the spatial agglomeration pattern shows a slow transition toward spatial dispersion.

Following the same analytical framework, the scatter plot of overall TSUE is shown in Fig. 6.18. During 1990–2020, most data points are concentrated in the High–High and Low–Low agglomeration quadrants, while the number of cities in the Low–High and High–Low outlier quadrants is relatively small and shows a declining trend. This indicates that the overall TSUE of individual cities has no significant difference from that of their neighboring cities, further confirming the existence of spatial autocorrelation in overall TSUE. In addition, from the evolutionary characteristics of the scatter plots, the spatial distribution of cities in terms of overall TSUE gradually shifted from dispersion to agglomeration during 1990–2020, which was particularly evident in the two periods of 1990–2000 and 2010–2020. It can be concluded that the agglomeration degree of overall TSUE has been gradually strengthened.

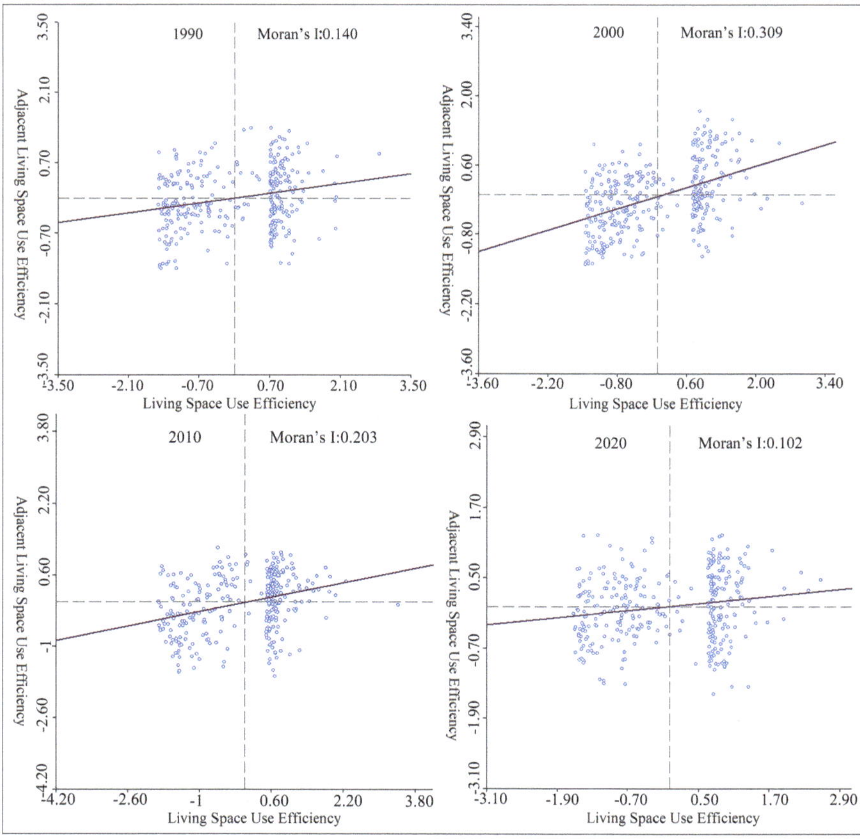

Fig. 6.16 Global Moran's I scatter plot of living space use efficiency

In summary, obvious positive spatial correlation and spatial agglomeration characteristics exist in PSUE, LSUE, ESUE, and overall TSUE across the 340 cities nationwide. The underlying reasons can be analyzed from two perspectives. First, according to geographical laws, geographical spaces and attributes are spatially correlated with one another. This study adopts city-scale territorial spaces as the basic DMUs for measuring TSUE. From the perspective of geographical mechanism, TSUE of adjacent cities inherently presents spatial correlation. Second, from the perspective of research connotation, territorial space use is a complex and interactive comprehensive process involving fundamental elements such as land, resources, population, and environment. The flow and exchange of these essential elements inevitably establish close connections among cities. Consequently, TSUE exhibits significant correlated characteristics in spatial distribution.

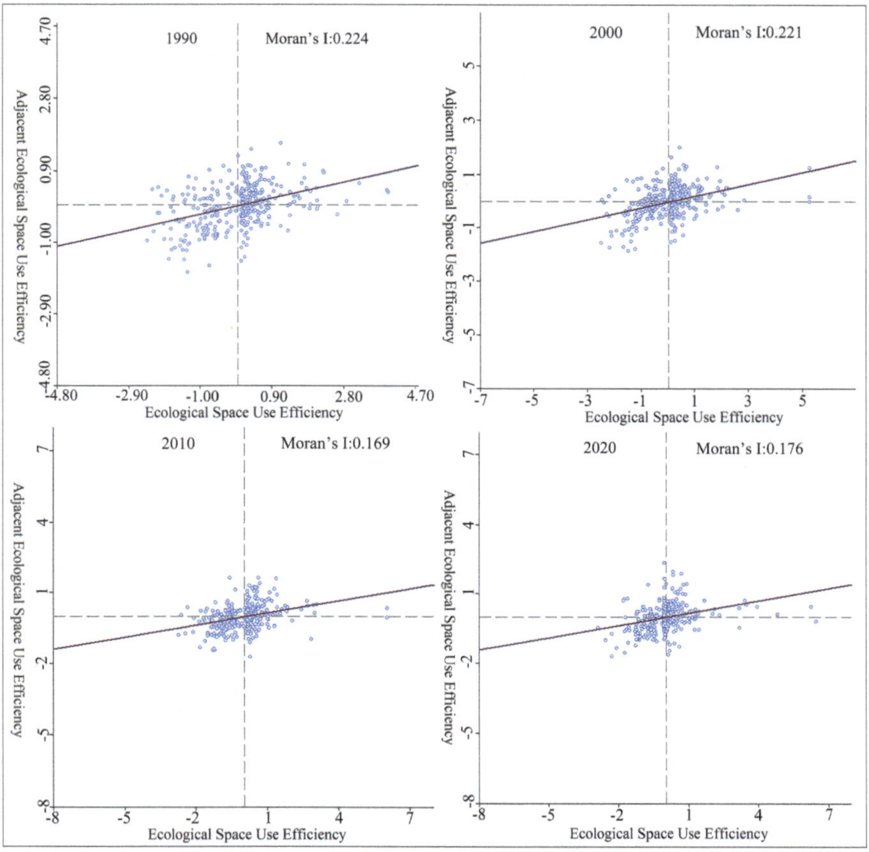

Fig. 6.17 Global Moran's I scatter plot of ecological space use efficiency

6.3.2.2 Local Spatial Autocorrelation of TSUE

Global spatial autocorrelation is an indicator that measures the spatial correlation of TSUE of 340 cities across China from an overall perspective. It mainly reflects whether the efficiency values show an aggregated or dispersed spatial trend. Such trend analysis helps grasp the overall correlation and potential inherent laws of TSUE at the macro level. Nevertheless, although global spatial autocorrelation provides a holistic analytical perspective, it cannot finely depict the specific differences and interactions among local regions. To further explore and reveal the local spatial correlation of TSUE, interpret the spatial interaction among cities, and clarify the specific location distribution and characteristics of local spatial agglomeration and anomaly, this study conducts cluster and outlier analysis of TSUE and generates corresponding

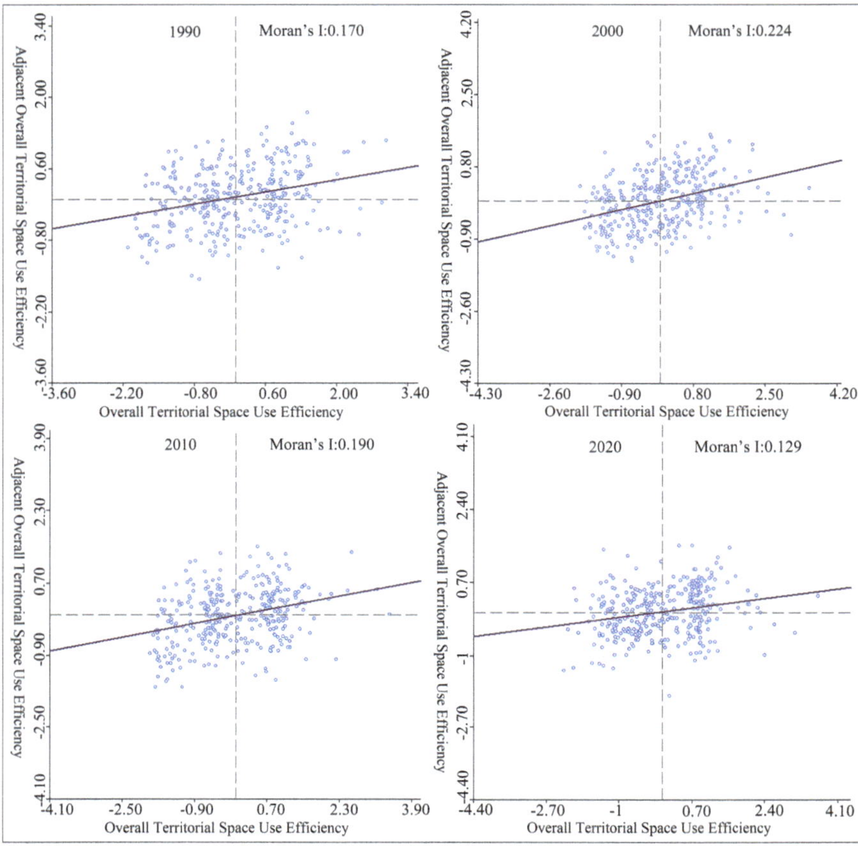

Fig. 6.18 Global Moran's I scatter plot of overall territorial space use efficiency

LISA maps with the support of ArcGIS Pro 3.0 and the local spatial autocorrelation model. This study attempts to identify agglomeration areas and outliers respectively for PSUE, LSUE, ESUE, and overall TSUE, followed by detailed analysis as presented below.

As shown in Fig. 6.19, the number of cities in agglomeration and outlier zones of PSUE changed markedly from 1990 to 2020, especially during the two periods of 2000–2010 and 2010–2020. From 2000 to 2010, the number of cities in H–H and L–L agglomeration zones decreased, showing an obvious weakening trend of agglomeration effects. From 2010 to 2020, the number of cities in agglomeration zones increased, accompanied by a corresponding strengthening of agglomeration effects. According to the analytical results in Sect. 6.3.1.1, the overall decline in PSUE in 2010 constituted a key turning point within the entire study period, indicating that changes in efficiency level directly affected the intensity of spatial agglomeration. Specifically, the overall drop in PSUE weakened the radiation and driving effect of advantageous cities, thereby influencing the spatial agglomeration of the whole region. The

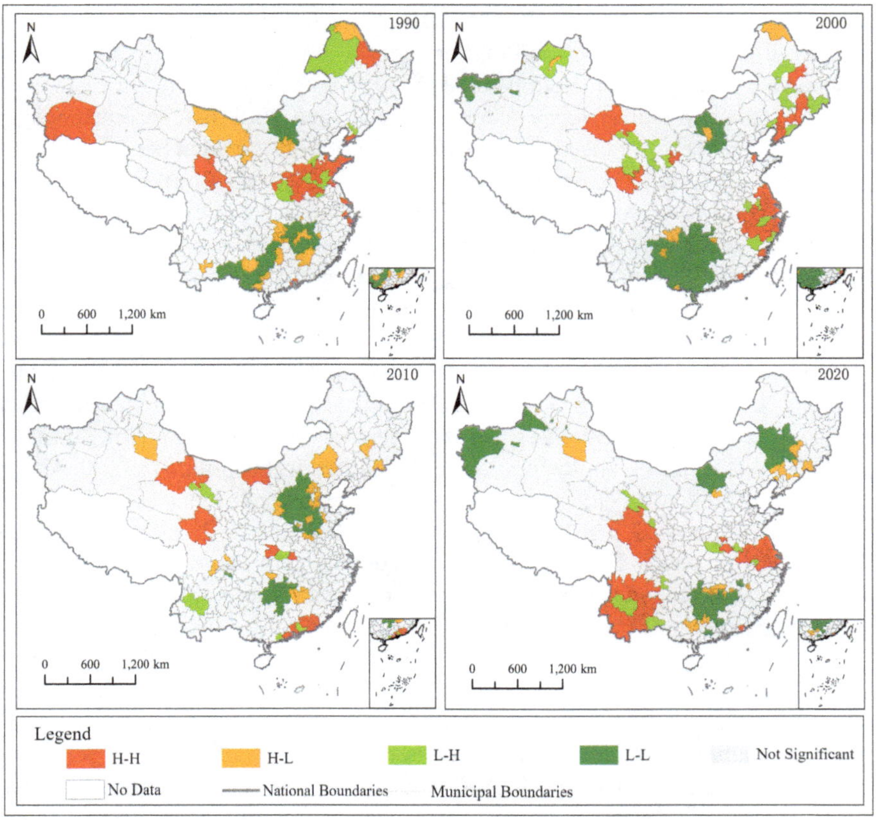

Fig. 6.19 LISA agglomeration map of production space use efficiency[12]

H–H agglomeration zones of PSUE are mainly concentrated in coastal provinces, including Shandong, Jiangsu, Zhejiang, and Guangdong. Endowed with superior geographical location, solid technological strength, and high socio-economic development level, cities in these regions and their surrounding areas generally maintain high PSUE. The L–L agglomeration zones are primarily distributed at the junction of Hunan, Guizhou, Shanxi, and Inner Mongolia, where both local and neighboring cities exhibit low PSUE, which deserves adequate attention from authorities. Relevant departments can improve the overall PSUE of this region by adjusting regional policies, strengthening scientific and technological innovation capacity, and learning from the development models of H–H agglomeration areas. H–L and L–H outlier zones are scattered sporadically around H–H and L–L agglomeration zones with relatively small quantities, showing substantial efficiency disparities with their surrounding areas.

[12] This figure is compiled based on the base map with the review approval number GS (2019) 1822.

As shown in Fig. 6.20, from 1990 to 2010, there were a relatively large number of cities in the agglomeration and outlier zones of LSUE with minor fluctuations, presenting a stable agglomeration trend. However, the number of cities in both agglomeration and outlier zones decreased significantly in 2020, accompanied by an obvious weakening of the agglomeration effect. As illustrated in Sect. 6.3.1.2, the overall level of LSUE remained relatively low in 2020, which further verifies that a low-efficiency level can affect spatial agglomeration to a certain extent. The H–H agglomeration zones of LSUE are mainly distributed in Hebei, Shandong, Jiangsu, Gansu and other provinces, with their agglomeration scale evolving over time. For instance, the scale of H–H agglomeration zones in Hebei and Shandong shrank noticeably from 1990 to 2000, while it tended to expand in 2010 and 2020 with the distribution range gradually shifting southward. The L–L agglomeration zones are largely concentrated in the Liaodong Peninsula, Hubei, Hunan, Henan and other provinces, where both local cities and their neighboring cities show low levels of LSUE. The scale of L–L agglomeration zones reached its maximum in 2000. As can be seen from Table 6.10, cities with low and medium–low efficiency accounted for a relatively large proportion in 2000, which explains the expanded agglomeration scale in that year. H–L and L–H outlier zones are scattered intermittently around H–H and L–L agglomeration zones in relatively small numbers. These areas feature a large efficiency gap with their surrounding regions and thus deserve adequate attention.

As shown in Fig. 6.21, from 1990 to 2020, the number of cities in the agglomeration and outlier zones of ESUE remains relatively large with small variation ranges, indicating a relatively stable agglomeration trend. The H–H agglomeration zones of ESUE are mainly distributed in the northwest and western regions, where both local and neighboring cities maintain high ESUE with evident positive spatial correlation. These regions possess superior original ecological background conditions and relatively slower socio-economic development compared with other areas, resulting in limited disturbance to the ecological environment and consequently a generally high level and prominent agglomeration of ESUE. The L–L agglomeration zones are primarily located in northern, northeastern China, and some provinces south of the Yangtze River. Their agglomeration scale changed over time. For example, the agglomeration zone centered on Hunan Province shrank markedly from 1990 to 2000, while its scale expanded significantly in 2010 and 2020. The number of cities in H–L and L–H outlier zones is small, with a wide and scattered distribution mainly across Inner Mongolia, Heilongjiang, Shaanxi, Hubei, and other regions. These areas show a substantial efficiency gap with their surroundings and therefore merit sufficient attention.

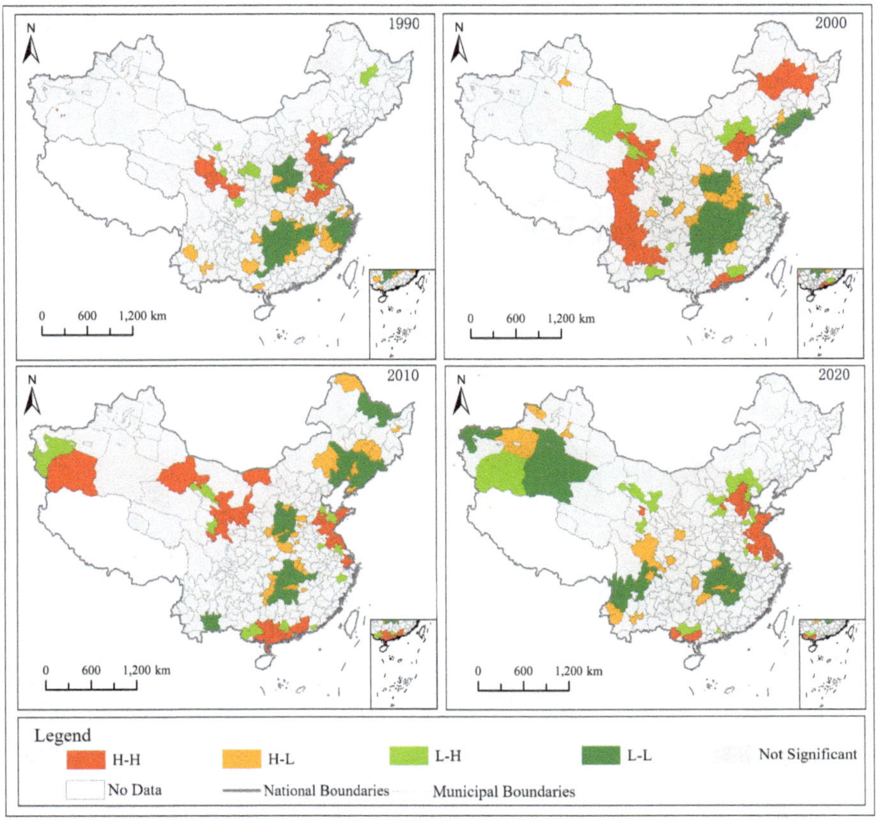

Fig. 6.20 LISA agglomeration map of living space use efficiency[13]

As shown in Fig. 6.22, the number of cities in the agglomeration and outlier zones of overall TSUE exhibited a continuous decreasing trend from 1990 to 2020, reaching the smallest quantity and the weakest agglomeration effect in 2020. As indicated in Table 6.12, the proportion of cities at low-efficiency and high-efficiency levels was relatively low in 2020, which partly explains the insignificant agglomeration effects of H–H and L–L types. The H–H agglomeration zones of overall TSUE are mainly concentrated in Shandong, Jiangsu, Guangdong, Gansu, and other regions. Cities in these areas maintain high-efficiency levels and meanwhile drive their neighboring cities to achieve high efficiency. The L–L agglomeration zones are primarily distributed in the Liaodong Peninsula, Hebei, Hunan, and other areas, where both local and surrounding cities present low-efficiency levels. The scale of L–L agglomeration zones kept changing over time. For instance, the contiguous agglomeration area covering Hunan, Hubei, and Guangdong expanded obviously

[13] This figure is compiled based on the base map with the review approval number GS (2019) 1822.

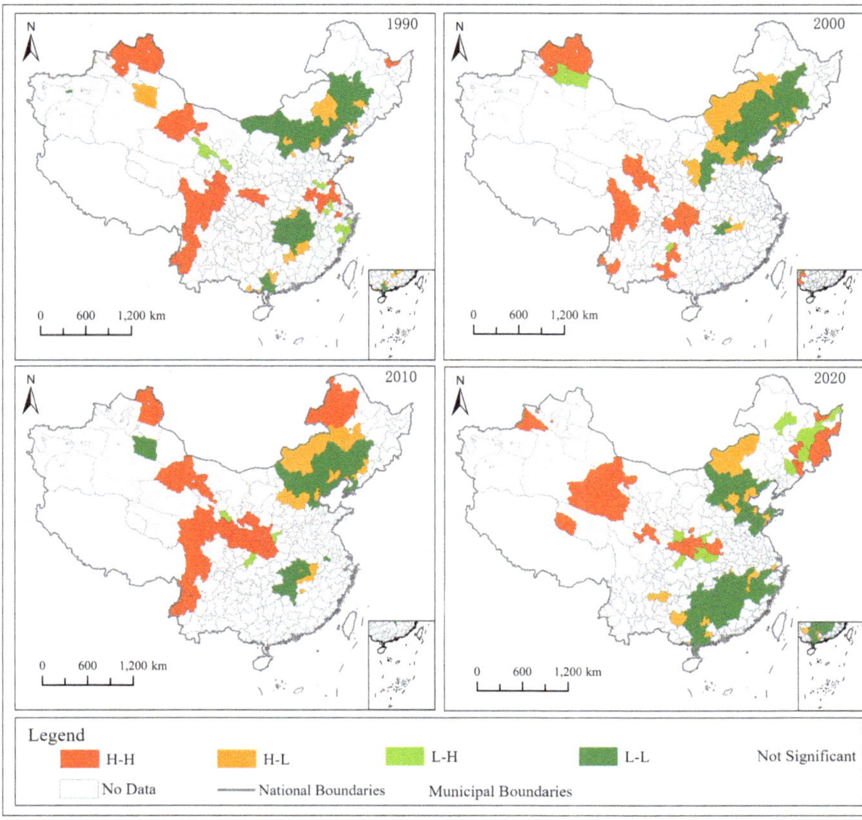

Fig. 6.21 LISA agglomeration map of ecological space use efficiency[14]

from 1990 to 2000 with a prominent agglomeration effect. By contrast, the agglomeration scale contracted in 2010 and 2020, accompanied by a gradual weakening of the agglomeration effect. Cities in H–L and L–H outlier zones are small in number and widely scattered around the agglomeration zones in Gansu, Liaoning, Guangdong, and Hunan. These regions show a distinct efficiency disparity with their surrounding areas and thus deserve adequate attention.

In summary, cities within H–H agglomeration zones can be regarded as diffusion centers. While maintaining a high-efficiency level themselves, they drive neighboring cities to achieve similarly high efficiency through spatial spillover effects. Therefore, these regions should maintain a reasonable input–output ratio to sustain their positive driving influence on surrounding cities. Cities in L–L agglomeration zones act as early warning lowland centers. They not only remain at a low-efficiency level themselves but also constrain the efficiency improvement of neighboring cities to a certain extent. Accordingly, such regions should be prioritized for targeted attention,

[14] This figure is compiled based on the base map with the review approval number GS (2019) 1822.

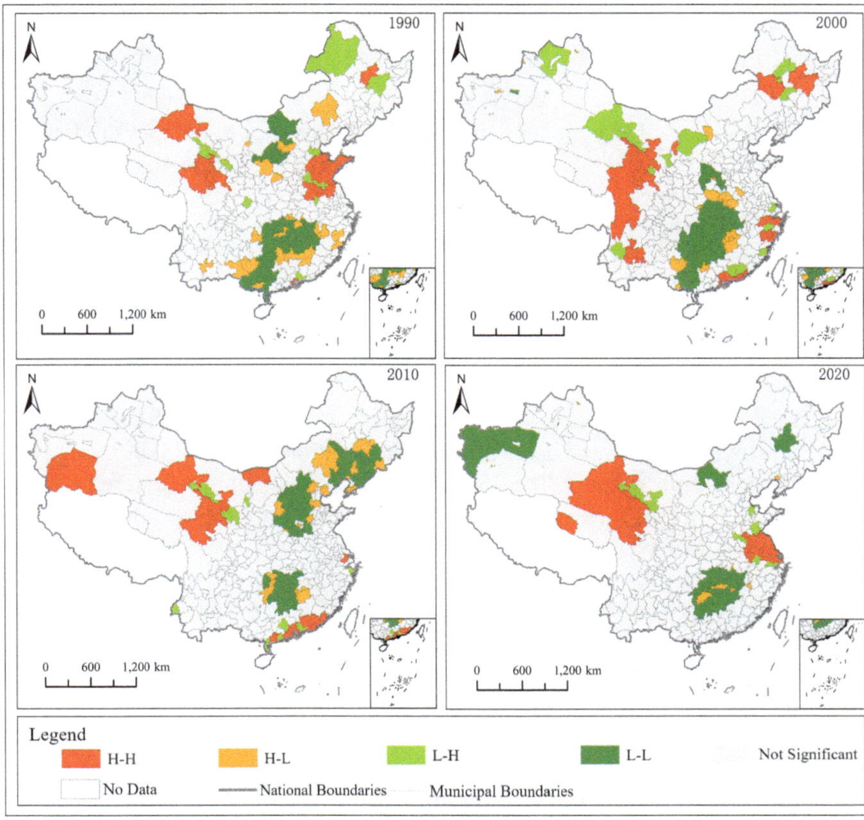

Fig. 6.22 LISA agglomeration map of overall territorial space use efficiency[15]

regulation, and governance. By developing competitive local industries, improving infrastructure, and protecting ecological environments in line with regional conditions, their TSUE can be effectively enhanced. H–L and L–H outlier zones represent polarization centers. In other words, cities with high (or low)-efficiency tend to inhibit (or promote) the efficiency development of their neighbors. Such areas require in-depth internal cause analysis. By means of policy guidance, industrial transformation, and strengthened interregional communication and cooperation, they can optimize their territorial space use patterns and achieve efficiency growth. Overall, according to their own efficiency level and the spatial correlation characteristics with neighboring cities, each city can implement targeted territorial space governance and optimization, so as to improve TSUE across the entire region.

[15] This figure is compiled based on the base map with the review approval number GS (2019) 1822.

6.4 Prediction and Analysis of Territorial Space Use Efficiency in 2030

Based on the extended CA model constructed in Chapter 5, this paper predicts the PSUE, LSUE, ESUE, and overall TSUE of 340 cities in 2030. The efficiency classification method adopted in Sect. 6.3 is further applied to conduct classified statistics on the prediction results.

6.4.1 Temporal Variation of TSUE in 2030

From a temporal perspective, as shown in Table 6.13, the overall TSUE presents a steady and slight upward trend in the future. The variation range of PSUE is relatively small. This is mainly because China's economic development has entered a stable stage of the new normal, with the growth rate shifting from high-speed to medium–high speed, and the development model transforming from scale-and-speed oriented to quality-and-efficiency oriented. Consequently, production activities no longer rely on increased input to expand output and accelerate economic growth, keeping PSUE rising at a mild pace. Compared with 2020, the number of cities at low, medium–low and medium-efficiency levels decreases slightly by 3, 5, and 2, respectively; meanwhile, the number of cities at medium–high and high-efficiency levels increases correspondingly by 5 each. In addition, as China enters the stage of high-quality development, greater attention has been paid to living and ecological spaces, and their spatial patterns have been optimized and adjusted through infrastructure construction and policy protection. Accordingly, the LSUE and ESUE undergo relatively obvious changes during 2020–2030. In terms of LSUE, the number of cities at low- and medium-efficiency levels changes dramatically, with the former dropping sharply by 54 and the latter surging by 51, while the quantity of cities at other efficiency levels changes slightly. As for ESUE, the number of cities at low and medium–low efficiency levels fluctuates marginally. The number of cities at the medium-efficiency level decreases significantly from 196 to 159, with its proportion falling by 10.89 percentage points. The number of medium–high efficiency cities rises evidently from 94 to 123, with a proportional increase of 8.53 percentage points, and the number of high-efficiency cities grows from 15 to 20. From 2020 to 2030, the overall TSUE shows a steady mild growth trend. The number of low-efficiency cities remains unchanged; the number of medium–low and medium-efficiency cities decreases by 11 and 2, respectively; the number of medium–high and high-efficiency cities increases by 9 and 4, respectively.

Table 6.13 Statistics on the number of cities at different levels of overall territorial space use efficiency in 2030 (Unit: number)

	Low-efficiency level	Medium–low efficiency level	Medium-efficiency level	Medium–high efficiency level	High-efficiency level
PSUE	33(9.71%)	69(20.29%)	144(42.35%)	76(22.35%)	18(5.29%)
LSUE	42(12.35%)	70(20.59%)	146(42.94%)	72(21.18%)	10(2.94%)
ESUE	4(1.18%)	34(10.00%)	159(46.76%)	123(36.18%)	20(5.88%)
Overall TSUE	11(3.24%)	108(31.76%)	143(42.06%)	67(19.71%)	11(3.24%)

6.4.2 *Spatial Pattern of TSUE in 2030*

From a spatial perspective, this study produces graded spatial visualization of PSUE, LSUE, ESUE, and overall TSUE in 2030, as shown in Fig. 6.23. Overall, the spatial patterns of PSUE, LSUE, ESUE, and overall TSUE in 2030 still present prominent unbalanced development with complex and diversified characteristics.

The spatial distribution of PSUE indicates that low-efficiency levels are concentrated in Northeast China, north-central Inner Mongolia, and western Xinjiang. As a key region under the national economic development strategy, the Northeast China has experienced economic recession due to geographical location, climatic conditions, historical background, industrial transformation, population changes, and other factors. To a certain extent, this results in its low PSUE in 2030. Northeast China is in urgent need of economic revitalization. It can foster high-quality development and sustainable revitalization, and improve PSUE by developing competitive industries, enabling technological empowerment, and consolidating strengths while addressing weaknesses. Some cities in north-central Inner Mongolia are adjacent to Northeast China and are similarly affected in economic development, leading to low PSUE. Industrial transformation and innovation should be prioritized in future development to comprehensively enhance PSUE. Western Xinjiang features a sparse population, weak economic foundation and relatively backward development, resulting in low PSUE. In the future, local agriculture, tourism, energy, and other industries should be developed based on local resource endowments to promote the rational utilization of production space. In addition, the central, coastal, southern, and southwestern regions maintain relatively stable economic development and achieve medium–high PSUE.

In terms of the spatial distribution of LSUE, cities with low and medium–low efficiency are mainly located in Northeast China and Yunnan Province. Due to its geographical location, Northeast China has a cold climate and long winters, which exert notable impacts on residential life. Moreover, statistical data reveal severe population loss and an increasingly severe aging problem in Northeast China, imposing adverse effects on the utilization of living space. Therefore, population revitalization is pivotal to the revitalization of Northeast China. It is necessary to strengthen talent

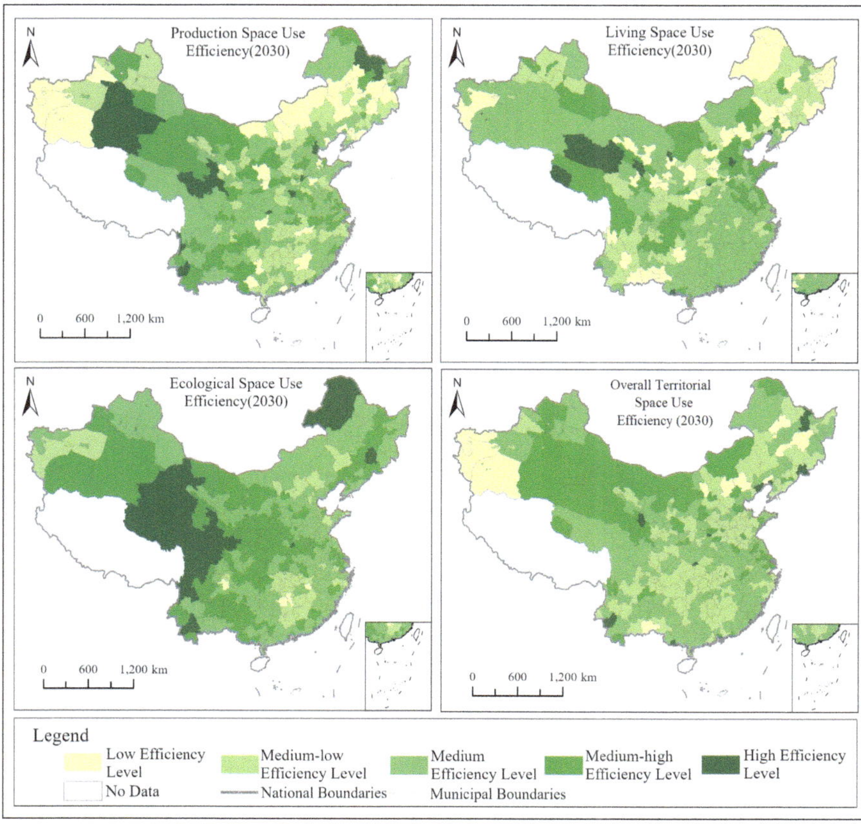

Fig. 6.23 Spatial distribution of territorial space use efficiency in 2030[16]

introduction, improve infrastructure such as education, medical care, and leisure facilities, and boost LSUE. Yunnan once had the largest number of poverty-stricken counties in China. Although poverty alleviation has been achieved, the utilization level of living space and residents' living standards remain at a relatively basic stage. Hence, persistent efforts are required to increase income and raise public living standards as well as LSUE in future development. Cities in other regions maintain medium or higher LSUE, which indirectly indicates a promising prospect of stable and happy lives for residents in the future.

The spatial distribution of ESUE shows that cities with low-efficiency and medium–low efficiency are primarily concentrated at the junction of Hubei, Hunan, and Jiangxi provinces. These regions have a high urbanization rate, developed economy, and dense population, and human production and living activities impose substantial negative impacts on the ecological environment, resulting in lower ESUE compared with other areas. In the future, it is essential to further reduce pollutant

---

[16] This figure is compiled based on the base map with the review approval number GS (2019) 1822.

emissions, enhance the connectivity of ecological space, and strictly protect ecological red lines to improve ESUE. Additionally, ESUE remains at a relatively high level in Northwest China, Northeast China, Henan, Shaanxi, Chongqing, and other regions.

The spatial distribution of overall TSUE presents a complex and diverse pattern. High-efficiency cities are sporadically distributed in Heilongjiang, Jilin, Liaoning, Qinghai, Guangxi, and other provinces without obvious agglomeration characteristics. It is evident that the driving role of urban agglomerations and economic belts in China needs to be further strengthened to boost the development and expansion of economic clusters, living circles, and ecological circles. Cities in eastern Xinjiang, western Inner Mongolia, Qinghai, Gansu, Jiangsu, and other provinces achieve medium–high efficiency overall TSUE and should maintain the sound utilization status of production, living, and ecological spaces. Prefecture-level cities in Shandong, Hubei, Hunan, Chongqing, and Jiangxi have relatively low overall TSUE. While pursuing economic development, these regions should attach importance to improving the use efficiency of production and living spaces to ensure the overall promotion of overall TSUE. Low-efficiency cities are scattered in parts of Heilongjiang, northern Hebei, southern Xinjiang, and other areas, where prominent problems exist in the utilization of production, living, and ecological spaces. It is necessary to deeply analyze the underlying causes and comprehensively promote the rational utilization and efficiency improvement of PLES.

In conclusion, TSUE is projected to maintain a steady upward trend by 2030. This prediction is not only consistent with the national strategic goals of territorial space optimization and efficient utilization but also inversely verifies the prediction accuracy of the extended CA model in TSUE simulation, strongly demonstrating the applicability and reliability of the model in the field of efficiency measurement.

6.5 Chapter Summary

6.5.1 Key Findings: National TSUE Features and Evolution Trends

This chapter adopts the theory of PLES. It integrates the GeoFA, GeoDEA and extended CA models. The research systematically evaluates TSUE across 340 Chinese cities from 1990 to 2020. It further analyzes the spatiotemporal evolution and spatial autocorrelation of efficiency values. It also predicts relevant efficiency levels for 2030. Core findings fall into four categories. These findings clearly reflect the overall characteristics and evolution trends of TSUE in China.

First, this study constructs a scientific and feasible evaluation indicator system of TSUE at the municipal scale. The GeoFA model conducts dimensionality reduction and screening on preliminary indicators. The final selection includes six input indicators and six output indicators for production space. It contains seven input indicators and six output indicators for living space. It also includes five input indicators and six output indicators for ecological space. The entropy weight method determines the weight value of production space at 0.4431. The weight value of living space is 0.4272 and that of ecological space is 0.1297. The comprehensive value of TSUE is calculated accordingly. The research achieves accurate quantification of multi-dimensional efficiency and comprehensive efficiency. It lays a solid indicator foundation for subsequent academic research.

Second, the PSUE, LSUE, ESUE, and overall TSUE all present distinctive temporal characteristics and obvious spatial imbalance. The year 2010 marks a critical turning point in temporal evolution. PSUE experiences a fluctuation trend of growth followed by decline and subsequent recovery. LSUE undergoes a change process of decline followed by growth and subsequent decline. ESUE maintains a steady growth trend. The changing trend of comprehensive efficiency remains consistent with that of PSUE. Efficiency values decline during the extensive economic development period from 2000 to 2010. These values rebound gradually after 2010 under the guidance of high-quality development. Spatial patterns show clear regional differentiation. PSUE maintains a high level in central and eastern China and a low level in northern, western, and southern China. LSUE stays high in northwest and northern China and remains low in parts of northeast, central, and southwest China. ESUE presents a pattern of high values in western and southern regions and low values in eastern and northern regions. Comprehensive efficiency generally keeps a high level in eastern coastal areas and a low level in parts of northern, central, and western regions. Low-efficiency cities and high-efficiency cities show an interlaced spatial distribution.

Third, TSUE of 340 cities across China shows significant positive spatial correlation and obvious agglomeration features. All Global Moran's I indices pass the significance test at the 1% level with positive numerical values. High-efficiency cities tend to gather with neighboring cities of high efficiency. Low-efficiency cities tend to cluster with surrounding cities of low efficiency. Local spatial autocorrelation results reflect clear regional differentiation. H–H agglomeration areas are mainly distributed in eastern coastal zones and ecologically superior northwest regions. These areas exert positive spillover and driving effects on surrounding regions. L–L agglomeration areas concentrate in parts of northeast central and southwest China. These regions form low-value areas requiring early warning and targeted regulation. H–L and L–H outlier areas scatter across different regions. These areas show prominent efficiency gaps with surrounding territories.

Fourth, TSUE is projected to achieve steady and moderate growth by 2030. PSUE maintains a stable development state. The number of low-efficiency cities in living space decreases substantially, and the number of medium-efficiency cities increases remarkably. The quantity of medium–high efficiency and high-efficiency cities in ecological space rises noticeably. The number of medium–high efficiency and high-efficiency cities in terms of comprehensive efficiency increases steadily. The overall development trend conforms to the national strategic orientation of TSUE.

Spatial imbalance will still exist in the future development process. Key optimization measures are still necessary for typical regions.

6.5.2 Framework Discussion: Practical Value of the Analytical System

This chapter strictly follows the overall analytical framework of "policy and data support—model construction—empirical analysis" established in Sect. 6.1.3. Taking the TSUE at the municipal scale of 340 cities across China as the research object, it integrates multiple research methods including the GeoFA, GeoDEA, extended CA models, and spatial autocorrelation analysis. A complete closed-loop research framework of "preliminary preparation—model construction—empirical verification—future prediction" is formed. It not only achieves the core research goals of "measurement—evolution—prediction", but also highlights the adaptability and application value of the framework. The methodological reflections summarized from practical application can also provide a reference for similar studies.

The core advantage of the framework lies in its scientifically designed logical closed loop, which effectively addresses the difficulties of multi-dimensional, long-time series and cross-scale research on TSUE, and realizes the organic connection of all research links. Policy and data support constitute the fundamental foundation. From the policy perspective, national strategies such as ecological civilization construction and territorial spatial planning are sorted out to clarify the research orientation and ensure the research results conform to practical demands. From the data perspective, multi-source data are integrated, and standardized preprocessing is adopted to solve the problems of data heterogeneity and inconsistent statistical caliber, providing high-quality support for subsequent research. Model construction connects three major tasks: indicator screening, efficiency measurement, and future prediction. Empirical analysis verifies the effectiveness of the model through temporal-spatial evolution and spatial autocorrelation analysis, forming a closed-loop framework of "method—data—application" with both theoretical depth and practical significance.

The collaborative application of multiple methods within the framework improves the scientificity and accuracy of the research. In the indicator screening stage, the GeoFA model eliminates indicator redundancy and multicollinearity to screen out core multi-dimensional indicators. In the efficiency measurement stage, the GeoDEA model breaks through the limitations of traditional models, incorporates geographical characteristics, and accurately depicts the input–output relationship of various spatial efficiencies. In the future prediction stage, the extended CA model realizes the scientific simulation of the efficiency pattern in 2030. In the spatial evolution analysis stage, spatial autocorrelation analysis reveals the spatial agglomeration characteristics of efficiency, making up for the deficiencies of traditional time-series analysis.

Meanwhile, there is still room for optimization in the framework application. Limited accessibility to some long-time series data forces the adoption of data from adjacent years, which may affect measurement accuracy. The collaboration depth of models, parameter transmission, and scenario coupling need to be further optimized. The universality of the framework remains to be tested and improved through applications in multi-scale and multi-region research. In conclusion, the framework is well adapted to municipal-scale research demands. Its application value and optimization direction can provide an important methodological reference for similar studies and help improve the methodological system of TSUE research.

Competing Interests The authors have no conflicts of interest to declare that are relevant to the content of this chapter.

References

1. Liu DY, Zheng XQ, Wang HB et al (2018) Interoperable scenario simulation of land-use policy for Beijing-Tianjin-Hebei region, China. Land Use Policy 75:155–165
2. Liu JL, Liu YS, Li YR (2017) Classification evaluation and spatial-temporal analysis of "production-living-ecological" spaces in China. Acta Geogr Sin 72(07):1290–1304
3. Chuai XW, Huang XJ, Wang WJ (2015) Land use, total carbon emission's change and low carbon land management in Coastal Jiangsu, China. J Clean Prod 103:77–86
4. Rong TQ, Zhang PY, Zhu HR et al (2022) Spatial correlation evolution and prediction scenario of land use carbon emissions in China. Eco Inform 71:101802
5. Lai L (2011) Carbon emission effect of land use in China. Dissertation, Nanjing University
6. Ma Y, Zheng MR, Xu F et al (2023) Modeling and exploring the coordination relationship between green infrastructure and land use eco-efficiency: an urban agglomeration perspective. Environ Sci Pollut Res 30(40):92537–92554
7. Ma Y, Zheng XQ, Liu ML et al (2022) Spatio-temporal evolution characteristics analysis and optimization prediction of urban green infrastructure: a case study of Beijing, China. Sci Rep 12(1):10702
8. Xie GD, Zhang CX, Zhang LM (2015) Improvement of the evaluation method for ecosystem service value based on per unit area. J Nat Resour 30(08):1243–1254
9. Zhong JT, Zhang XY, Gui K et al (2022) Reconstructing 6-hourly PM datasets from 1960 to 2020 in China. Earth System Science Data 14(7):3197–3211

Chapter 7
Policy Recommendations for Improving China's Territorial Space Use Efficiency

Abstract This chapter proposes policy suggestions to improve China's TSUE. Relying on three theoretical underpinnings including spatial heterogeneity, spatial correlation and efficiency's dynamic evolution, it completes the logical shift from efficiency measurement to spatial governance. Targeted differentiated pathways are formulated for PLES according to their respective constraint conditions. Furthermore, the study categorizes all regions into four categories and matches customized supporting policies correspondingly. From the perspectives of property rights, market mechanisms, planning control and digital governance, this research puts forward feasible institutional safeguards and innovations of policy instruments. The constructed policy framework highlights differentiated arrangement and holistic coordination, which can provide systematic support for the high-quality development of territorial space in China.

7.1 Theoretical Foundation: From Efficiency Measurement to Spatial Governance

Scientific research on TSUE ultimately serves governance practice. It completes the logical transition from efficiency measurement to spatial governance.

This transition rests on three theoretical pillars. The first is the spatial heterogeneity theory of efficiency. Measurement results derived from the GeoDEA model reveal obvious differences in efficiency level and structure across cities and regions. These disparities stem from the divergence of resource endowments and development stages. They also indicate differentiated paths for efficiency improvement and targeted space for policy intervention. The second is the spatial correlation theory of efficiency. Spatial autocorrelation analysis verifies that high-efficiency cities and low-efficiency cities do not exist independently. These features shape regional efficiency patterns via spatial spillover effects. Efficiency promotion policies for a single city must take regional coordination into account. This mechanism can avoid zero-sum competition with neighboring regions. The third is the dynamic evolution theory of efficiency. Prediction outputs of the extended CA model present potential scenarios

Y. Ma, *Evaluation and Prediction Modeling of Territorial Space Use Efficiency*, Urban Sustainability, https://doi.org/10.1007/978-981-92-4396-9_7

of TSUE in 2030. Such forward-looking cognition provides a time window for proactive policy intervention. It also promotes the shift of policy formulation from ex post response to ex ante regulation. On the basis of the above theories, efficient governance needs to establish a closed loop management system covering identification, diagnosis, intervention, and evaluation. Technical outputs of the GeoFA, GeoDEA, and extended CA models can be converted into operable policy instruments.

7.2 Multidimensional Optimization Paths: Differentiated Strategies for Production-Living-Ecological Spaces

Based on the independent measurement and comparative analysis of TSUE for PLES in Chap. 6, this section proposes differentiated improvement pathways according to the restrictive factors and promotion potential of efficiency in each dimension.

The pathway for PSUE improvement focuses on industrial structure optimization and intensive spatial utilization. The GeoDEA model measurement identifies low-efficiency areas of production space, which are mainly distributed in resource-based cities of central and western China and old industrial bases in Northeast China. Policy priorities lie in industrial transformation and spatial restructuring. Industrial restructuring can eliminate high-energy-consumption and low-output industries and introduce high-tech manufacturing and modern service industries to improve unit land output efficiency. The promotion of multi-story standard factory buildings and vertical development modes can increase the plot ratio of industrial land and ease the conflict between industrial land expansion and cultivated land protection. For high-efficiency regions of production space such as the Yangtze River Delta and Pearl River Delta urban agglomerations, policy emphasis shifts to efficiency maintenance and radiational driving. Mechanisms such as industrial enclaves and joint park construction can spread technology, capital, and management experience to surrounding low-efficiency areas and restrain the excessive widening of regional efficiency gaps.

The pathway for LSUE improvement lays emphasis on the equalization of public services and the optimization of job-housing spatial layout. The measurement results of LSUE reveal the coexistence of urban malaise in large cities and insufficient public services in small and medium-sized cities. Relevant policies should exert efforts from both supply and demand sides. On the supply side, a public service allocation mechanism linked to permanent population should be established to break restrictions on the household registration system on the distribution of educational, medical, and other public resources. Node cities within urban agglomerations with continuous population inflow predicted by the extended CA model need reasonable advance layout of schools, hospitals, and elderly care facilities. On the demand side, the transit-oriented development model can optimize urban spatial structure, shorten commuting distance, and reduce spatial friction costs caused by job-residence mismatch. Cities with persistently low LSUE should prioritize the improvement of community living circles and enhance the accessibility and satisfaction of basic public services.

The pathway for ESUE improvement highlights the balance between value realization of ecological products and protective utilization. The measurement of ESUE reflects the misconception that ecological protection inevitably leads to low efficiency. Scientific evaluation criteria for ecological efficiency should be formulated to incorporate carbon sequestration function, biodiversity conservation, and landscape recreational value into the output accounting of ecological spaces. Such arrangement forms an incentive mechanism that rewards effective protection and imposes costs on ecological damage. Specific measures include the improvement of ecological compensation mechanisms. Compensation standards and transfer payment scales can be determined according to the ecological service supply capacity measured by the GeoDEA model. Exploration of value realization approaches for ecological products can develop ecological agriculture, health tourism, and carbon sink trading and transform ecological advantages into developmental advantages. The strict delimitation and enforcement of ecological protection red lines are also essential. The prediction function of the extended CA model can identify high-risk areas of ecological space occupation and support the establishment of early warning and rapid response mechanisms.

7.3 Zonal Classified Policy Framework: Governance Strategies for Spatial Efficiency Patterns

Combined with the efficiency levels and spatial clustering results measured by the GeoDEA model, as well as the 2030 efficiency evolution trend predicted by the extended CA model, this section constructs a three-dimensional policy matrix covering efficiency type, development stage, and spatial location to support targeted policy implementation.

High-efficiency leading areas mainly refer to core cities of urban agglomerations along eastern coastal regions. Their policy orientation lies in efficiency maintenance and model demonstration. These regions have reached the efficiency frontier. Policy focus shifts from quantitative expansion to quality improvement. The scale of newly added construction land shall be strictly controlled to promote the renewal and redevelopment of stock space. An efficiency monitoring and early warning system can be established to prevent efficiency decline caused by industrial relocation and population aging. Institutional and practical experience of efficiency improvement can be summarized and refined into replicable and popularizable policy tools for external promotion through regional cooperation mechanisms.

Efficiency catching-up areas are dominated by provincial capital agglomerations and regional central cities in central and western China. Their policy orientation focuses on efficiency leapfrogging and path breakthrough. Such regions maintain moderate efficiency levels with great growth potential. Relevant policies should seize the development window in the middle and late stage of urbanization. Territorial space development patterns can be optimized to guide the agglomeration of population and industries in superior areas and generate scale effects. Increased investment in infrastructure and public services can remedy historical infrastructure deficiencies and improve the hardware environment for efficiency growth. Innovative policies concerning land, fiscal taxation, and industry can attract high-efficiency industries and talents and facilitate the transformation and upgrading of development momentum.

Efficiency potential areas mostly include small and medium-sized cities and special-type regions in central and western China. Their policy orientation emphasizes efficiency cultivation and bottom-line constraint policies. These regions present low-efficiency levels and insufficient driving forces for improvement. Policy formulation should balance efficiency and equity. Central fiscal transfer payment and paired assistance can be strengthened for resource-depleted cities, border cities, and other special-type regions to guarantee basic public service supply. These measures can also avoid excessive population loss and social recession triggered by efficiency competition. Ecologically vulnerable areas shall strictly implement the major function zoning system and restrain large-scale industrialization and urbanization. Local authorities can explore eco-oriented, green, and low-carbon development paths and give priority to ESUE.

Efficiency recession risk areas are identified as declining-efficiency regions based on the simulation results of the extended CA model. Their policy orientation focuses on proactive intervention and transitional support. Some cities in Northeast China and resource-depleted cities may face efficiency stagnation or decline according to model simulation. An early warning mechanism for efficiency recession can be set up to formulate advance plans for industrial transformation and population loss response. Special support funds can be arranged to support urban function reshaping and stock space revitalization. Local governments can explore the optimization and adjustment of administrative divisions. Urban amalgamation and functional area restructuring can expand effective development space and improve the allocation efficiency of regional resources.

7.4 Institutional Safeguards and Policy Tool Innovation

The effective implementation of efficiency improvement pathways relies on the optimization of institutional environment and the innovation of policy instruments. A supporting system can be established from four dimensions, including property right system, market mechanism, planning regulation, and digital governance.

Innovation of the property right system serves as the fundamental guarantee for efficiency improvement. The property right system of natural resource assets should be improved, and the unified right verification and registration of natural resources should be promoted. This mechanism clarifies property right subjects and interest boundaries of various resources within territorial space. A unified urban and rural construction land market can be established to break institutional barriers that lead to unequal rights and prices between collective land and state-owned land. The reform enables the market to play a decisive role in resource allocation. Mechanisms for the transfer and transaction of development rights under territorial space use regulation can be explored. High-efficiency regions are allowed to purchase construction land indicators from low-efficiency regions, so as to balance efficiency gains and regional equity.

The improvement of market mechanisms acts as the driving source of efficiency improvement. A pricing formation mechanism that reflects market supply and demand as well as resource scarcity should be perfected. Setting higher entry thresholds and use cost of industrial land can restrain inefficient spatial expansion. The value realization mechanism of ecological products needs further improvement. A systematic framework for the accounting and application of ecosystem service values can be established to facilitate the trading of ecological products such as forest and wetland carbon sinks and clean energy green certificates. The development of a multi-level capital market can support high-efficiency enterprises to integrate inefficient land through mergers and reorganizations and promote the optimal allocation of spatial resources.

The optimization of planning regulation provides spatial guarantee for efficiency improvement. An efficiency-oriented territorial spatial planning system should be constructed. Measurement results of the GeoDEA model and prediction scenarios of the extended CA model can be incorporated into planning formulation references to improve scientific rationality and forward-looking capacity of planning schemes. Dynamic evaluation and adjustment mechanisms for planning implementation should be refined. Regular efficiency evaluation can offer timely early warnings for regions deviating from planning objectives and initiate planning revision procedures. Planning law enforcement and supervision can be strengthened with remote sensing monitoring and big data technology. Real-time monitoring on territorial space utilization helps impose strict sanctions against illegal construction and inefficient land use.

Digital governance empowerment offers technical support for efficiency improvement. A digital platform for the monitoring and evaluation of TSUE can be built. The platform integrates algorithm frameworks of the GeoFA, GeoDEA, and extended CA models and realizes integrated and automatic operation in indicator screening, efficiency measurement, and trend prediction. The construction of a basic information platform for territorial space can be advanced to break inter-departmental data barriers. Real-time data sharing and business collaboration can be achieved across natural resources, ecological environment, economic, and social sectors. Intelligent auxiliary decision-making systems can be explored. Efficiency big data and machine learning algorithms can deliver intelligent and scenario-based decision support for policy formulation.

7.5 Chapter Summary

This chapter completes the logical closed loop from academic research on efficiency to the design of governance policies. Based on the technical outputs of the GeoFA, GeoDEA, and extended CA models, this study constructs a policy framework for the improvement of TSUE featuring multi-dimensional, zonal, and phased arrangements. It puts forward differentiated improvement pathways for PLES. It divides the study area into four policy zones according to efficiency patterns and clarifies the development orientation and corresponding strategies of each zone. It also innovates institutional guarantees and policy instruments from four dimensions covering property right, market mechanism, planning regulation, and digital governance. The established policy framework embodies the governance philosophy of adopting tailored measures in line with local conditions and classified implementation. It emphasizes overall coordination between efficiency promotion and regional development, ecological protection, as well as social equity. It provides systematic policy guidance for the high-quality development of territorial space in the new era. Future research can further deepen the effect evaluation and dynamic optimization mechanism of efficiency-related policies. It can explore the iterative feedback between model algorithms and policy practice and continuously improve the refined and intelligent governance level of territorial space governance.

Competing Interests

The authors have no conflicts of interest to declare that are relevant to the content of this chapter.

Kolibri 360 GmbH
Ferdinand-Jühlke-Straße 7
99095 Erfurt, Deutschland
produktsicherheit@kolibri360.de